Applications of Fluidization to Food Processing

Applications of Fluidization to Food Processing

P. G. Smith

University of Lincoln
UK

Blackwell
Science

© 2007 by Peter G. Smith

Blackwell Science, a Blackwell Publishing company
Editorial offices:
Blackwell Science Ltd, 9600 Garsington Road, Oxford OX4 2DQ, UK
Tel: +44 (0)1865 776868
Blackwell Publishing Professional, 2121 State Avenue, Ames, Iowa 50014-8300, USA
Tel: +1 515 292 0140
Blackwell Publishing Asia Pty Ltd, 550 Swanston Street, Carlton, Victoria 3053, Australia
Tel: +61 (0)3 8359 1011

First published 2007

ISBN: 978-0-632-06456-4

Library of Congress Cataloging-in-Publication Data

Smith, P.G.
Applications of fluidization to food processing / P.G. Smith. – 1st ed.
p.   cm.
Includes bibliographical references and index.
ISBN-13: 978-0-632-06456-4 (hardback : alk. paper)
ISBN-10: 0-632-06456-0 (hardback : alk. paper)
1. Fluidization.   2. Fluid dynamics.   3. Food industry and trade.   I. Title.
TP156.F65S65 2007
664'.024–dc22

2006102874

A catalogue record for this title is available from the British Library

Set in 10 on 13 pt Palatino
by SNP Best-set Typesetter Ltd., Hong Kong
Printed and bound in Singapore
by Utopia Press Pte Ltd

The publisher's policy is to use permanent paper from mills that operate a sustainable
forestry policy, and which has been manufactured from pulp processed using acid-free
and elementary chlorine-free practices. Furthermore, the publisher ensures that the text
paper and cover board used have met acceptable environmental accreditation standards.

For further information on Blackwell Publishing, visit our website:
www.blackwellpublishing.com

For Liz, Alexis, Imogen and Verity

Contents

Preface

Fluidization is a versatile technique for bringing about intimate contact between finely divided solids and a fluid. Although its origins are a little obscure, the first significant industrial application of fluidization was the catalytic cracking of heavy hydrocarbons using beds of catalyst particles in the 1940s. The extension to a wide range of other chemical, especially exothermic, reactions came about because heat can be transferred relatively easily to and from the bed and bed temperatures are uniform and can be held constant. The application of fluidization to a number of physical unit operations followed this early exploitation in reaction engineering. Very considerable research was undertaken in universities and research institutes, and in industry, between the 1960s and 1980s when much of the fundamental behaviour of gas-solid fluidized beds in particular was established and there followed a vast literature dealing with bubbles, mixing, heat and mass transfer, and so on. Very many fluidization research groups have been active in chemical engineering departments throughout the UK, USA, Australia, Canada, the then USSR and most European countries. In the UK this work was headed by, amongst many others, Rowe at UCL, Davidson at Cambridge, Richardson at Swansea and Geldart at Bradford. The lead in understanding the phenomenon of spouted beds was taken by Epstein and Mathur in Vancouver.

Although the relevance of fluidization to food processing was recognised at an early stage, there is little in the literature which is specific to the fluidization of food particulates. On the one hand, the subject is poorly treated in almost all food technology and food engineering textbooks and, on the other, much of the specialised fluidization literature can be a little unapproachable to those with a specific interest in food applications. Topics such as fluidized bed freezing are simply ignored in the chemical engineering literature. This book is an attempt to bridge this gap. Much of what is written about foods and fluidization appears to have the food element 'tacked on' and if this appears also to be true of the present work then it is because of the general paucity of relevant literature. It is difficult, but not impossible, to make experimental measurements on food systems in fluidized beds but it seems that few researchers endeavour to do so. Equally, it must be said that the fundamentals of fluidization are often misunderstood; fluidization is a very powerful processing technique but it cannot solve every solids processing problem and it is not appropriate in every situation. This lack of relevant food-related fluidization literature has made it necessary to rely upon relevant work in other fields. For example, in Chapter 5 there is much to be gained in understanding the mechanisms of fluidized bed granulation by examining work with materials as diverse as nuclear waste and fertiliser.

It has not been possible to cover all aspects of the principles of fluidization. A number of comprehensive texts on fluidized bed behaviour are available and inevitably I have drawn heavily on these. The reader who wishes to go into greater depth about the fundamental mechanisms at work in fluidized beds should consult those works by Davidson and Harrison (1971), Botterill (1975), Davidson, Clift and Harrison (1985), Kunii and Levenspiel (1991) and more recently Gibilaro (2001). Full references can be found at the end of Chapter 1. In addition, I have concentrated on gas-solid fluidized beds somewhat to the exclusion of liquid-solid fluidization although an indication of how particulate fluidization can be applied to biochemical reactors is given in Chapter 7.

Part One of the book is a description of fluidized bed behaviour and covers the theory necessary to understand the applications which follow. Chapter 1 is a description of the characteristics of both gas-solid and liquid-solid fluidized beds, concentrating especially upon the predication of minimum fluidizing velocity. Chapter 2 covers the all-important topics of heat transfer, mass transfer and particle mixing which underpin almost all the applications of fluidization, which are treated in Part Two.

Over the last 40 years freezing has grown in importance because the quality of the frozen product is often significantly better than that produced using more traditional thermal preservation methods. Fluidized bed freezing (Chapter 3) finds particular application in the processing of large-volume products such as prepared vegetables and a variety of soft fruits. Despite the popularity of frozen food, however, drying remains a very widespread operation in the food industry. Fluidized bed driers (Chapter 4) exploit the rapid particle mixing and high heat transfer coefficients in aggregative fluidization and are one of the most versatile and widely used kinds of drier. The manufacture of particles with specific properties is of increasing importance in food processing; Chapter 5 covers the mechanisms of fluidized bed granulation and its application to agglomeration, layering, coating, instantising and encapsulation. Chapter 6 describes recent work on the use of gas-solid fluidized beds as novel bioreactors for fermentation reactions and the culture of micro-organisms. This technique has considerable advantages over conventional fermentation systems and has the potential to manufacture a range of volatile products, including for example flavour compounds, using a variety of micro-organisms. Chapter 7 covers, in rather less detail, a wide range of other uses for gas-solid fluidization, notably blanching, roasting, explosion puffing, sterilisation and atmospheric pressure freeze drying. The use of liquid-solid fluidization as a turbulence promoter in ultrafiltration and in sterilisation is outlined, as well as its use as a bioreactor where the liquid phase is the substrate and the solid phase usually takes the form of immobilised enzyme beads.

Finally, a brief note on spelling and terminology: the word 'fluidization' may also be spelt 'fluidisation'. The former has been used throughout this book except in the references to published works where the original published spelling has been retained. The term 'fluid bed' is also in common usage and in most cases is taken to mean the same as fluidized bed. Again, I have used this where it appears in the titles of books and papers but not otherwise.

I must express my sincere thanks to Dr Bill Hayes for permission to use material, including a number of diagrams, from his PhD thesis, which was written under my supervision. This appears in Chapter 6. My thanks also to Nigel Balmforth of Blackwell for his patience and encouragement over far too great a period.

P.G. Smith
Lincoln
October 2006

Glossary

Aggregative fluidization	Fluidization characterised by the formation of bubbles of fluidizing fluid. This usually occurs where the fluidizing medium is a gas.
Bubble phase	The discontinuous phase of an aggregative fluidized bed made up of bubbles of fluid (almost always gas).
Bubbling fluidization	*See* aggregative fluidization.
Carry over	Entrained particles which leave the fluidized bed column entirely.
Channelling	Above the nominal minimum fluidizing velocity, the passage of gas through narrow channels in a bed of solids; the bed may remain defluidized.
Cloud	At bubble rise velocities greater than minimum fluidizing velocity, the approximately spherical shell of gas which surrounds a bubble and circulates through it as it rises through the bed.
Dense phase	Continuous (i.e. the non-bubble) phase in an aggregative fluidized bed.
Dilute phase	Fluidization at high gas velocities where all particles are carried in the gas stream. This corresponds to pneumatic conveying.
Disperse phase	The region of decreasing solids concentration occupying the freeboard.
Distributor	The porous plate or grid through which the fluidizing fluid passes into the bed.

Elutriation	The fractionation or preferential separation of entrained particles which changes with distance up through the column above the bed surface.
Emulsion phase	*See* dense phase.
Entrainment	The transport of particles into the exhaust gas stream above the bed surface.
Fast fluidization	Fluidization of a bed of moderate solids concentration formed at extreme gas velocities by the recycling to the fluidized column of transported particles.
Fixed bed	*See* packed bed.
Fluid bed	An alternative term for fluidized bed.
Fluidized bed	A bed of particulate solids through which a fluid passes, thus imparting liquid-like properties to the solids.
Freeboard	The space between the surface of the dense phase and the gas exit.
Geldart classification	A system of classifying particles into four groups (A, B, C and D) according to their fluidization behaviour, first proposed by Geldart.
Heterogeneous fluidization	*See* aggregative fluidization.
Homogeneous fluidization	*See* particulate fluidization.
Incipient fluidization	*See* minimum fluidization.
Lean phase	*See* bubble phase.
Minimum fluidization	The state forming the boundary between a fixed bed and a fluidized bed. The point at which the drag force on a particle is equal to its net weight.
Minimum fluidizing velocity	The superficial gas velocity at and beyond which the bed is said to be fluidized.
Packed bed	A stationary bed of particles up through which a gas or a liquid passes at low fluid velocities.
Particulate fluidization	Fluidization characterised by continuous and uniform bed expansion. In general, this occurs

	when the fluidizing medium is a liquid.
Slugging	A state (axial slugging), at high superficial gas velocities, beyond the bubbling bed, where bubbles grow to the size of the column and are known as slugs. In an alternative type of slugging the bed is divided into alternate rising regions of dense phase and disperse phase.
Spout	The trail of particles drawn up by a bubble as it rises through a fluidized bed.
Spouted bed	A bed of relatively large particles (usually over 1 mm in diameter), with a conical gas inlet, resulting in a regular circulatory pattern with a central high-velocity spout and an annulus of slow downward-moving particles.
Superficial velocity	The volumetric flow rate of fluidizing fluid divided by the cross-sectional area of the bed.
Turbulent fluidization	At very high superficial gas velocities, discrete bubbles or slugs are no longer present and the bed is characterised by significant pressure fluctuations.
Two-phase theory	The assumption that all gas over and above that required for minimum fluidization flows up through the bed in the form of bubbles.
Voidage	Properly, interparticle voidage; the fraction of a bed of particles occupied by free space.
Wake	The space between the indented base of a bubble and the bubble sphere. It is occupied by particles as a bubble rises in the bed.

Part One

Fundamentals of Fluidization

Chapter 1
A Description of Fluidized Bed Behaviour

An introduction to fluidization

Consider a bed of particulate solids or powder, say of a size similar to table salt. When a fluid, either a gas or a liquid, is passed upwards through the bed, the bed particles remain stationary or packed at low fluid velocities. This is a packed or fixed bed. If now the velocity of the fluid is increased, the particles will begin to separate and move away from one another; the bed is said to expand. On increasing the velocity further, a point will be reached at which the drag force exerted by the fluid on a particle is balanced by the net weight of the particle. The particles are now suspended in the upward-moving stream of fluid. This is the point of minimum fluidization, or incipient fluidization, at and beyond which the bed is said to be fluidized.

The superficial fluid velocity in the bed at the point of incipient fluidization is called the minimum fluidizing velocity u_{mf} and is of crucial importance in characterising the behaviour of a fluidized bed. Superficial velocity is defined as the volumetric flow rate of fluid divided by the cross-sectional area of the bed. In other words, superficial velocity is equal to the mean fluid velocity assuming that the particles are not present; the actual interstitial fluid velocity will, of course, be somewhat higher. At velocities in excess of that required for minimum fluidization one of two phenomena will occur. First, the bed may continue to expand and the particles space themselves uniformly. This is known as particulate, or homogeneous, fluidization (Figure 1.1) and in general occurs when the fluidizing medium is a liquid. Alternatively, the excess fluid, beyond that required to achieve fluidization, may pass through the bed in the form of bubbles. The particles are agitated and mixed, violently so at higher velocities, and the bubbles rise through the bed to break at the surface. This is called aggregative, bubbling or heterogeneous fluidization and usually occurs where the fluidizing medium is a gas. It is this type of behaviour which gives rise to the analogy of a boiling liquid (Figure 1.2).

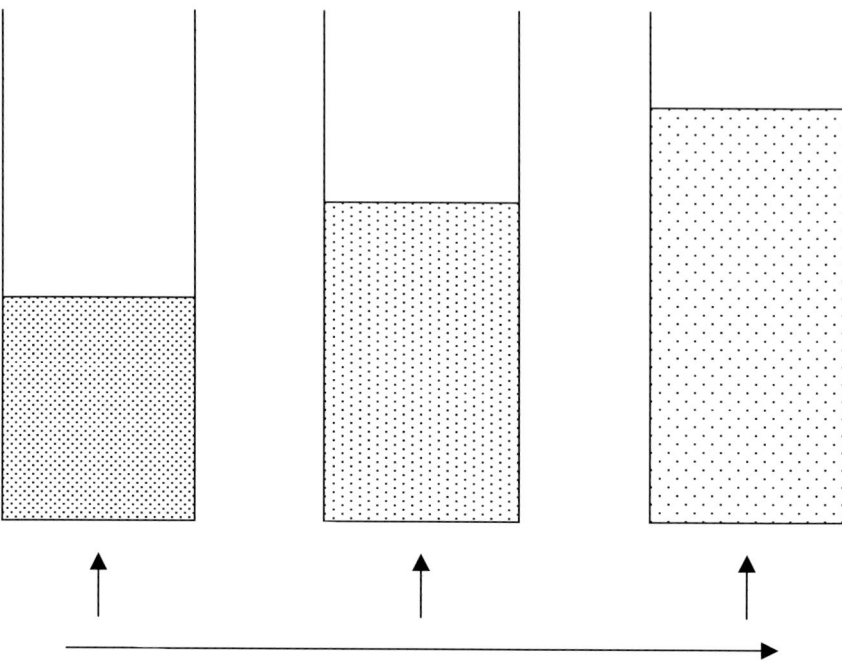

Increasing superficial velocity

Figure 1.1   Particulate fluidization.

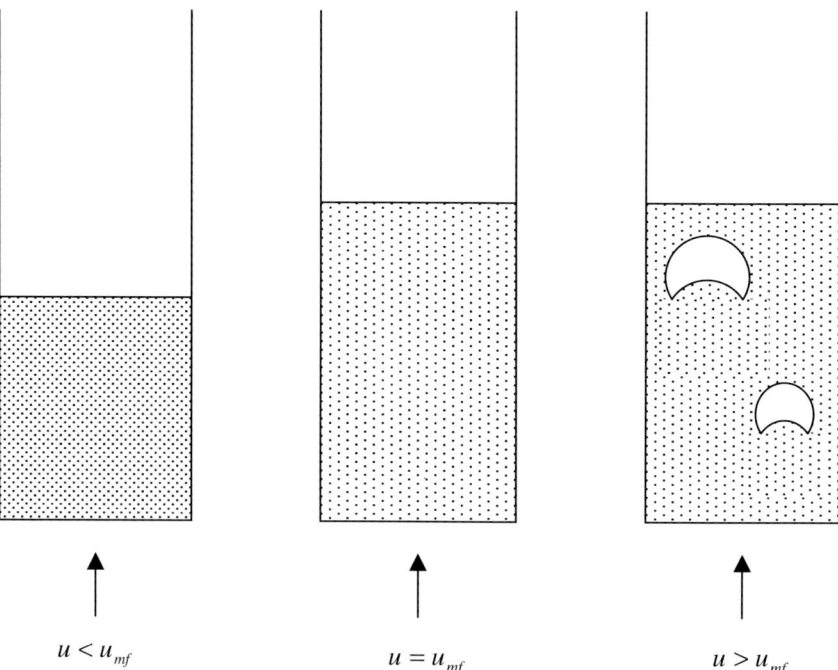

$u < u_{mf}$ $u = u_{mf}$ $u > u_{mf}$

Figure 1.2   Aggregative fluidization.

Thus fluidization is a technique which enables solid particles to take on some of the properties of a fluid. For example, solids fluidized by a gas will adopt the shape of the container in which they are held and can be made to flow, under pressure, from an orifice or overflow a weir. If the wall of the bed is punctured by a series of apertures aligned vertically the fluidized solids will behave just as if the bed were filled with liquid; a stream of solids will issue from each aperture, that from the highest point in the bed will travel only a short horizontal distance whereas the stream from the lowest aperture will travel furthest. Gibilaro (2001) refers to a demonstration rig in the Department of Chemical Engineering at UCL in which a plastic toy duck buried in a bed of sand exchanges place with a brass duck on the bed surface when the bed is fluidized with air.*

It is possible to operate a fluidized bed in either batch or continuous mode. Strictly, most batch applications are in fact operated in semi-batch mode where the solids are treated as a batch but the fluidizing medium enters and leaves the bed continuously. In the case of gas-solid beds used in fermentation (see Chapter 6), the fluidizing gas is recirculated although reactants and products flow continuously. In true continuous operation the solids may be fed into a fluidized bed via screw conveyors, weigh feeders or pneumatic conveying lines and can be withdrawn from the bed via standpipes or by flowing over weirs.

Essentially aggregative fluidization is a two-phase system: there is a dense phase (sometimes referred to as the emulsion phase), which is continuous, and a discontinuous phase called the lean or bubble phase. The simplified assumption that all the gas over and above that required for minimum fluidization flows up through the bed in the form of bubbles is known as the two-phase theory. If the total volumetric flow of gas is Q then

$$Q = Q_{mf} + Q_B \qquad\qquad 1.1$$

where Q_{mf} is the volumetric flow of gas at minimum fluidization and Q_B is the volumetric flow rate of gas in the bubbles. The interstitial gas flows up through the dense phase of the bed in streamline flow. The particle Reynolds number (defined by equation 1.42) has a value between unity and about 10 for most fine particles, such as spray-dried

---

* I recall attempting to impress prospective undergraduates with this demonstration when I was a research student at UCL. From memory, rather than ducks, there were two plastic fish, one blue and weighted with sand and one pink. The demonstration purported to show how fluidized beds were able to change the colour of plastic fish . . . !

food powders, but increases by two or three orders of magnitude for large particles such as vegetable pieces in a fluidized bed freezer.

However, the rather idealised picture of a fluidized bed which has been described so far is not observed in all cases. There can be very different patterns of behaviour across a range of particle and fluid properties. For example, some solids are capable of being fluidized far more easily than others; Richardson (1971) suggests that a well-fluidized system is likely to result if the particles are fine, of low density, with an approximately spherical shape and narrow size distribution, and when the fluid is of high density. Although gas-solid systems normally give rise to aggregative fluidization and liquid-solid systems to particulate behaviour, in extreme cases particulate fluidization can occur with very small particles of low density which are fluidized by a dense gas at high pressure. In such cases increasing the gas velocity beyond u_{mf} results in significant bed expansion before bubbling begins and a minimum bubbling velocity u_{mb} may be defined as the gas velocity at which bubbles first appear. In the idealised two-phase theory, however, $u_{mf} = u_{mb}$.

Wilhelm and Kwauk (1948), in one of the earliest papers on fluidization, suggested that the Froude number, written for the minimum fluidizing condition, could be used to indicate the prevalent fluidization behaviour. Thus

$$Fr = \frac{u_{mf}^2}{gd} \qquad\qquad 1.2$$

where d is the mean diameter of the bed particles. Values of Fr below unity suggest particulate fluidization and values greater than unity suggest aggregative behaviour.

At velocities above those at which bubbling fluidization occurs, the nature of the contact between gas and solids changes significantly. As the gas velocity increases, especially in deep beds, the bubbles grow to the size of the bed container and push plugs of material up the bed as they rise. This condition is known as slugging, in which the bed particles stream past the slugs at the bed wall on their downward path. Beyond slugging, when the terminal falling velocity of the particles is exceeded, particles are entrained in the gas stream and considerable elutriation occurs. At this point large discrete bubbles are absent (Yerushalmi and Avidan, 1985) and the bed is said to be turbulent. Kunii and Levenspiel (1991) define a dense-phase bed as one in which there is a reasonably clearly defined bed surface, whether the fluidizing medium is gas or liquid. Further increase in the gas velocity beyond the turbulent fluidized condition leads ultimately to lean-phase fluidization and pneumatic conveying of the solids. However, if particles which have been elutriated are recycled to the bed at a sufficiently high rate then a relatively high concentration of solids can be maintained in the bed. This is the condition of fast fluidization (Yerushalmi and Avidan, 1985).

Although the majority of food applications exploit the characteristics of either the dense-phase beds or particulate beds which have been described so far, there are food applications of two other phenomena: spouted beds and centrifugal fluidization. A spouted bed is a method of allowing intimate gas-solid contact for larger particles, with a minimum diameter of perhaps 1 or 2 mm, and a more regular circulation pattern than is observed in aggregative fluidization (Mathur and Epstein, 1974). Spouted beds are covered in more detail in the section on Other Types of Fluidization, below. In the centrifugal fluidized bed (see section below entitled Centrifugal Fluidization) particles enter a perforated horizontal cylinder which rotates inside a plenum with air blown across the outside of the cylinder and perpendicular to the axis of rotation. The perforated surface of the cylinder acts as the distributor plate and the centrifugal action keeps the particle bed in place close to the cylinder wall. Its particular advantage is that fluidization requires higher gas velocities than normal, thus allowing, for example, significant heat input to the bed at lower temperatures than would otherwise be possible.

A number of brief but comprehensive descriptions of basic fluidized bed behaviour are available (Richardson, 1971; Botterill, 1975; Couderc, 1985) which are recommended as an introduction to the subject.

Industrial applications of fluidization

The phenomena of rapid particle movement and the intimate contact between solids and at least a portion of the gas give rise to a series of characteristics of aggregative fluidization such as good mixing, near isothermal conditions and high rates of heat and mass transfer which are exploited in a wide range of unit operations.

The first significant industrial use of fluidization was in the catalytic cracking of hydrocarbons in the 1940s and arguably the most widespread use today is as a reactor for heterogeneous catalytic reactions such as the production of acrylonitrile, maleic anhydride, low-density polythene and phthalic anhydride, amongst others (Yates, 1983; Kunii and Levenspiel, 1991). Other applications have included the calcination of uranyl nitrate and radioactive aluminium nitrate wastes (Jonke *et al.*, 1957), coal gasification, adsorption, drying of particulate solids (Kunii and Levenspiel, 1991), mixing (Geldart, 1992) and granulation (Sherrington and Oliver, 1981; Smith and Nienow, 1983) including the agglomeration of pharmaceutical powders (Rankell *et al.*, 1964). Epstein

(2003) has reviewed the uses of liquid-fluidized beds and these include particle classification, washing, adsorption and ion exchange, flocculation, electrolytic recovery of metals, bioreactions and liquid-fluidized bed heat exchangers where the fluidized particles enhance film heat transfer coefficients and reduce the fouling of heat transfer surfaces.

Applications of fluidization in the food industry

Minimum fluidizing velocity is approximately proportional to the square of particle diameter and to the difference in density between the particles and the fluidizing medium. The majority of the applications of aggregative fluidization in the chemical processing industries involve inorganic solids of relatively high density but particle sizes at most of the order of 1 or 2 mm. Jowitt (1977) suggested that the applications of fluidization in the food industry fall into two broad categories. In the first group, food pieces are fluidized directly and because the density of food materials, particularly vegetable matter, is only a little greater than that of water it is possible to fluidize relatively large particles such as potato chips, peas, sprouts and diced vegetables even with a gas as the fluidizing medium. Jowitt's second group of applications are those where a packaged food is placed in a fluidized bed of inert solids. The heating of sealed containers in this way (i.e. cans or jars in a sterilisation operation) has a number of advantages including improved thermal economy, better control of temperature and a reduction in the consumption of process water for cooling. In a development of this idea, Rios *et al.* (1985) described a 'fluidized flotation cell' in which larger lighter objects are fluidized in a bed of denser fine particles. Optimum operation occurs at object-to-bed density ratios between 0.5 and 0.9, object-to-bed volume ratios less than 0.4 and gas velocities two or three times greater than the minimum fluidizing velocity of the fines. Outside these conditions, the objects tend either to float on the bed surface or sink. Such systems also require separation of the floating objects from the fines; this is a possible disadvantage although the authors suggest that in some cases this can be accomplished easily by simple screening.

Many of the uses of gas-solid fluidized beds in the chemical and processing industries, for example as driers, mixers, granulators and reactors, are relevant also in food processing but in addition there are a number of food-specific applications such as freezing, blanching, cooking and roasting (Jowitt, 1977; Smith, 2003) and the sterilisation of canned foods (Jowitt, 1977). Shilton and Niranjan (1993) have reviewed the uses of fluidization in food processing. In fluidized bed freezing the solids to be frozen are fluidized by refrigerated air at temperatures of −30°C or below and the particles are frozen independently and very rapidly to give a free-flowing IQF (Individually Quick Frozen) product;

in a fluidized bed drier the enthalpy of vaporisation is supplied by the fluidizing gas which is usually air, although submerged heating elements can also be used; in mixing applications the bed particles are moved rapidly by the action of rising bubbles; granulation covers those processes which produce granules or instantised products from a solution or slurry and again the heat for evaporation of the solvent is usually supplied by the fluidizing gas. A related process is fluidized bed coating where the objective is to coat particles or tablets with a solid material such as lactose but avoid particle growth by agglomeration (Vinter, 1982).

Rios *et al.* (1985) reported the use of a fluidized bed for the roasting of coffee beans, resulting in improved quality compared to traditional rotating drum methods. However, the fluidized bed technique involved increased costs and higher levels of pollution due to the rejection of hot exhaust gases. Rios and co-workers have also proposed the use of a 'whirling bed apparatus' in which a wedge placed on the gas distributor plate induces a cyclical particle motion and high-intensity mixing (Rios *et al.*, 1985). Coupled with gas recycle, this experimental coffee roasting method allowed for aroma recovery and reduced energy consumption and gave an improved product quality. Roasting times were reduced from the 10–15 minutes in the conventional drum method to between 2 and 4 minutes in the whirling bed apparatus (Arjona *et al.*, 1980). The blanching of vegetables with a mixture of air and water vapour in a combined fluidized bed conveyor belt system has been reported to give a reduction in energy consumption of over 20% compared to water blanching, a significant reduction in the loss of vitamin C and reduced waste water treatment costs (Rios *et al.*, 1978, 1985).

Liquid fluidization is the basis of both the Oslo (or Krystal) continuous crystalliser (Mullin, 1993) which is used in the production of, for example, sugar or citric acid, and the bioreactors in which immobilised cells or enzymes are fluidized by the reactant solution (Epstein, 2003). It is used in the leaching of vegetable oils from seeds (Rios *et al.*, 1985; Epstein, 2003) and in physical operations such as the washing and preparation of vegetables.

The details of food processing applications are covered in Chapters 3–7; the remainder of this chapter deals with a detailed description of fluidized bed behaviour.

Gas-solid fluidized bed behaviour

Influence of gas velocity

The relationship between bed pressure drop and superficial fluidizing velocity is shown in Figure 1.3. As the gas velocity increases, the

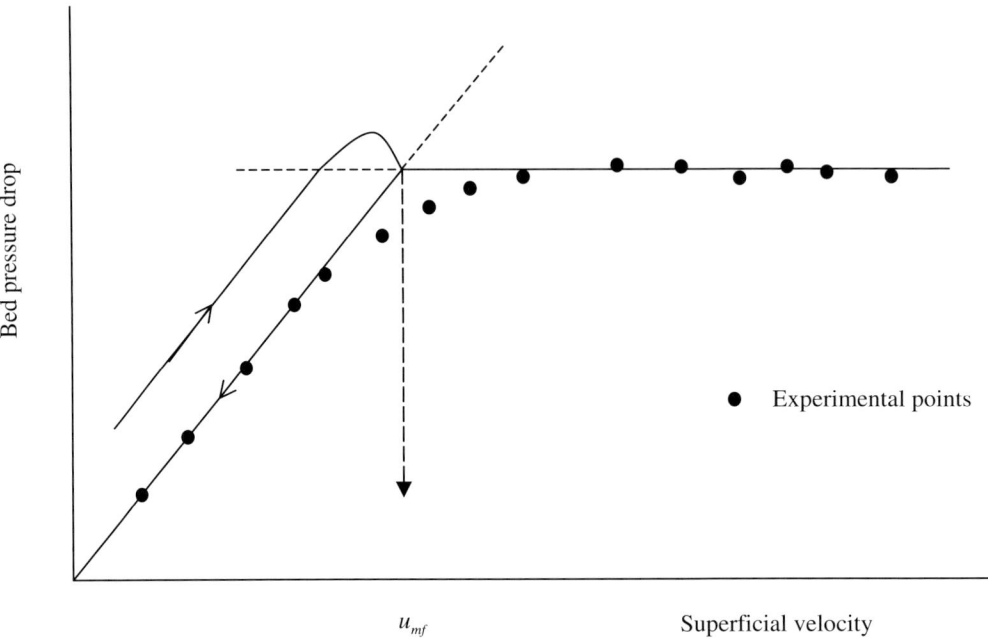

Figure 1.3 Relationship between bed pressure drop and superficial fluidizing velocity.

pressure drop increases in the fixed bed, or packed bed, region and then levels out as the bed becomes fluidized. Ideally, the pressure drop then remains constant as the weight of the particles is supported by the fluid. However, if the velocity is then reduced, marked hysteresis is observed. This is because the bed voidage remains at the minimum fluidizing value whereas with increasing gas velocity considerable vibration of the particles takes place, the voidage is lower and the pressure drop correspondingly is slightly greater. In practice, too, there will be a maximum pressure drop through which the curve passes because of particle interlocking. Also, as the velocity is reduced, the transition between the fluidized and fixed curves is gradual rather than sudden.

As the superficial gas velocity is increased beyond the minimum fluidizing velocity, a greater proportion of gas passes through the bed in the form of bubbles, the bubbles grow larger, particle movement is more rapid and there is a greater degree of 'turbulence'. Fluidizing gas velocity is the single most important variable affecting the behaviour of a bed of given particles and it is expressed usually as either:

(1) multiples of u_{mf}

e.g. $\dfrac{u}{u_{mf}} = 3$ implies that the gas velocity is three times that required for minimum fluidization, or as

(2) excess gas velocity, $u - u_{mf}$

e.g. $u - u_{mf} = 1.2\,\mathrm{m\,s^{-1}}$ implies that the gas velocity is $1.2\,\mathrm{m\,s^{-1}}$ *greater* than that required for minimum fluidization.

In practice, proportionately more gas flows interstitially (i.e. between the particles) as the velocity is increased than at u_{mf}. In addition, there is a limited interchange of gas between the bubble phase and the dense phase. As the gas velocity is increased further the very smallest particles are likely to be carried out of the bed in the exhaust stream. This is because at any realistic fluidizing gas velocity, the terminal falling velocity of the very smallest particles will be exceeded. The loss of bed material in this way is known as elutriation and will increase as $\dfrac{u}{u_{mf}}$ increases. Further increases in gas velocity result in greater elutriation and a more dilute concentration of the solids remaining in the bed. Eventually all the particles will be transported in the gas stream at the onset of pneumatic conveying.

The section above the fluidized bed surface is often referred to as the freeboard. This section may have a gradually increasing column diameter, rather like an inverted cone, which is designed to reduce the gas velocity and thus disengage gas and particles. Particles then fall back to a level where the superficial gas velocity is sufficient to support their weight.

Geldart's classification

Fluidized bed behaviour is affected not only by gas velocity but also by particle size and density. Based upon observations of fluidized bed behaviour in air at ambient conditions and at velocities below $\dfrac{u}{u_{mf}} = 10$, Geldart (1973) suggested classifying fluidized particles into four groups: C (cohesive), A (aeratable), B (sand-like) and D (spoutable). This classification is shown diagrammatically in Figure 1.4 in the form of a plot of the density difference between particle and fluid against mean particle size.

Group A particles are typically between $20\,\mu\mathrm{m}$ and $100\,\mu\mathrm{m}$ in diameter with a particle density less than $1400\,\mathrm{kg\,m^{-3}}$. These particles exhibit considerable bed expansion as the fluidizing velocity increases and collapse only slowly as the velocity is decreased. In other words, they tend to retain the fluidizing gas. The bubbles are limited in size and frequently split up and coalesce as they rise through the bed. On the other hand, the larger and denser group B particles ($40–500\,\mu\mathrm{m}$, particle density in the range $1400–4000\,\mathrm{kg\,m^{-3}}$) form freely bubbling fluidized beds at or slightly above the minimum fluidizing velocity. Small

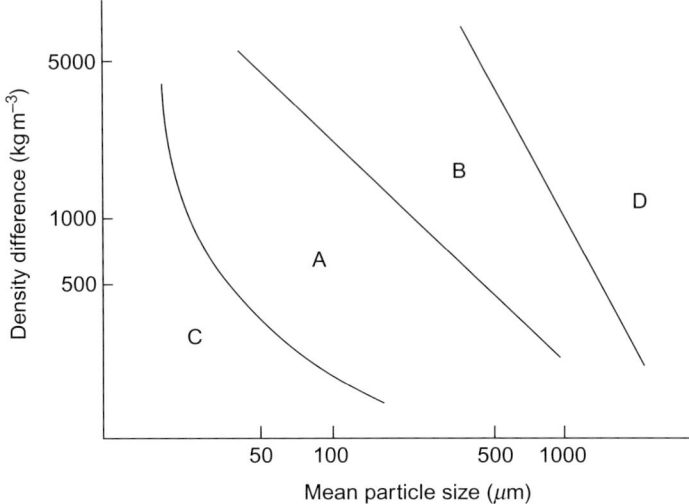

Figure 1.4   Geldart's classification.

bubbles form at the distributor and grow by vertical coalescence as they rise up the bed. The increase in bubble size is approximately linear with increasing excess gas velocity and height above the distributor plate and size is limited only by the diameter of the bed. Vigorous bubbling brings about considerable solids mixing. This is the classic fluidized bed behaviour; group B particles can be defined by

$$(\rho_s - \rho_f)^{1.17} \geq 9.06 \times 10^5 \qquad\qquad 1.3$$

Group C, with particle diameters below 30 μm, consists of cohesive powders which display a tendency to agglomerate and are very difficult to fluidize; the forces between particles are greater than the hydrodynamic force which is exerted on the particle by the fluidizing gas. In small-diameter beds the particles tend to move upwards as a solid plug. As the bed diameter increases, relatively little bubbling is observed and instead the gas passes up through the bed in a series of channels. This group is characteristic of a number of food materials such as flour or very fine spray-dried particles. Geldart (1992) suggested that these powders can be identified by measuring both the tapped and aerated bulk densities. If the ratio of the two exceeds a value of 1.4 then the particles are likely to be group C.

Large particles with a mean diameter greater than 600 μm and a density above 4000 kg m^{-3} are classed as group D and whereas in Geldart B beds the bubbles tend to rise faster than the gas in the dense phase, the reverse is true of bubbles in Geldart D particle beds. Bubbles tend to coalesce horizontally rather than vertically and reach a large size. The resulting solids mixing is poorer than in Geldart B beds. Slugging occurs if the bubble size approaches the bed diameter and large Geldart D beds are prone to spouting (see section on Spouted Beds below). Such particles display very little bed expansion and generally give unstable fluidization; channelling of the gas is prevalent. Food examples include seeds and vegetables pieces. Group D particles can be defined by

$$(\rho_s - \rho_f)d^2 \geq 10^9 \qquad\qquad 1.4$$

Bubbles and particle movement

Bubble formation at the distributor

The growth of a typical bubble at a small aperture in the distributor plate was described by Zenz (1971) and is illustrated in Figure 1.5 which shows the changes in the interface between the fluidizing gas entering the bed through an aperture in the distributor plate and the dense phase of the bed. The velocity u of gas passing through the aperture and into the bed will be greater than the minimum fluidizing velocity of the bed particles u_{mf} by at least an order of magnitude and therefore will be sufficient to lift the interface between gas and bed particles into the position shown in Figure 1.5b. As the void grows (Figure 1.5c), the mean velocity of gas passing through the interface will decrease by a factor equal to the ratio of the aperture area to the area of the interface. Zenz suggested that if this mean velocity is still greater than u_{mf} then the interface will remain stable and the void will grow still further, as in Figure 1.5c. At some point the void is large enough that the velocity through the interface decreases to the value of u_{mf} and therefore the void reaches its maximum size. The pressure exerted by the fluidized particles is greater at the base of the void than at its uppermost point, due simply to the depth of 'fluid'; that is, the depth of the dense phase. Thus the interface collapses at the base of the void, cutting off the incoming gas into a bubble with a more stable and near-spherical shape. A new interface between the inlet gas and bed particles is then formed and the process is continually repeated. Consequently the volumetric flow rate of gas and the minimum fluidizing velocity of the bed particles influence the initial size of bubbles formed at the distributor (Zenz, 1971).

Bubble growth and bubble shape

The bubbles in a fluidized bed have a distinct shape and there is a distinct boundary between the gas in the bubble and the

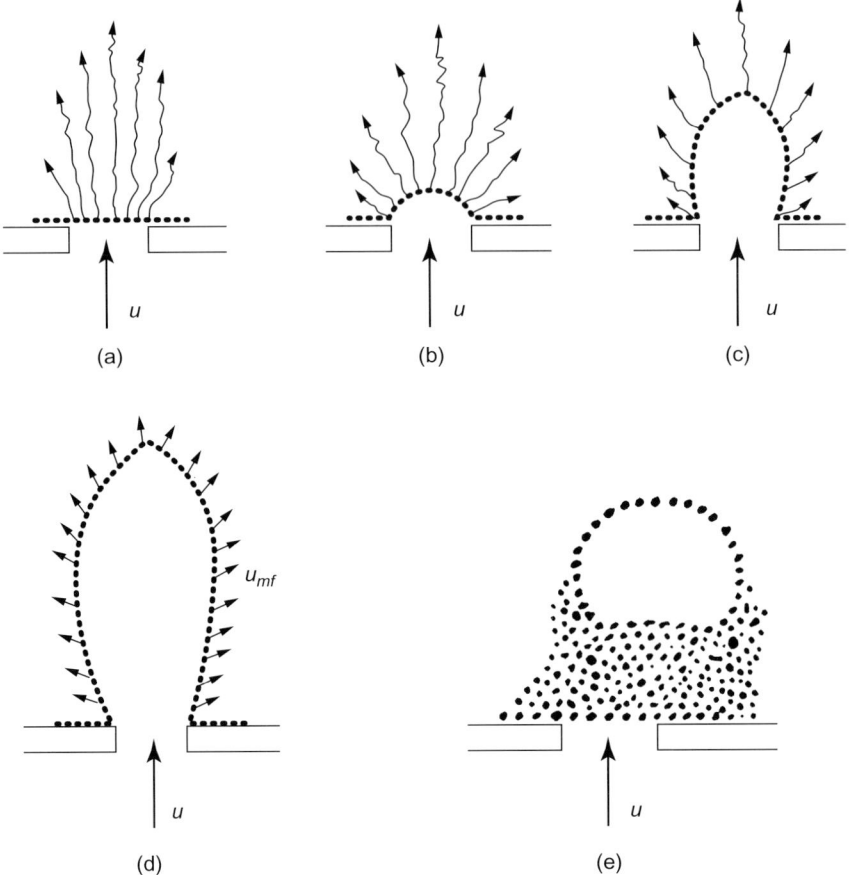

Figure 1.5   Bubble formation at the distributor. Reprinted from Davidson, J.F. and Harrison, D., Fluidization, Academic Press, 1971, with permission from Elsevier.

surrounding dense phase although the boundary tends to oscillate. A single bubble, rising in a bed where the bubble diameter is much less than the bed diameter, is essentially spherical but with a slight indentation at the base. However, the shape can be influenced and distorted by close proximity to either the bed walls and or submerged surfaces such as heat transfer coils or the nozzles used for injecting liquid in some applications. Both the shape and size are affected when the bubble diameter becomes greater than about half the bed diameter. Rowe (1971) suggested that the minimum size of a bubble is 1–2 orders of magnitude greater than the size of the particles in the bed, which implies a minimum diameter of about 0.005 m in fine powders and 0.10 m in beds of coarse particles with a diameter of several mm.

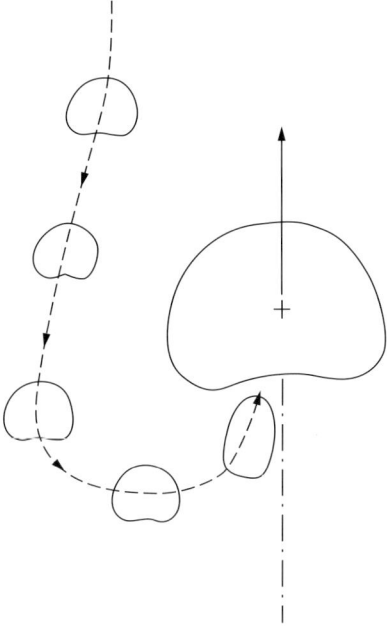

Figure 1.6 Coalescence of bubbles. Reprinted from Davidson, J.F. and Harrison, D., Fluidization, Academic Press, 1971, with permission from Elsevier.

Bubbles grow as they rise through the bed and, although there is some interchange of gas between bubbles and the dense phase, growth is due mainly to coalescence. Small bubbles are overtaken by larger ones, which rise at a greater velocity, and coalescence takes place at the base of the larger bubble, as in Figure 1.6. Vertical indentations often develop downwards from the roof of the bubble which then grow to split the bubble in half vertically but growth by coalescence is far more significant than any size reduction. As they grow, the size of the indentation at the base of a bubble increases and they take on the characteristic kidney-shape shown in Figure 1.7. The total volumetric flow rate of gas in the bubble phase is approximately constant and consequently the number of bubbles at any bed cross-section decreases with bed height.

As Rowe (1971) has shown in a series of experiments using a tracer gas in a two-dimensional bed, the interstitial gas in a bubbling bed enters the base of a bubble and emerges through the bubble roof, resulting in some interchange of gas between the dense and bubble phases. However, if the rise velocity of a bubble exceeds u_{mf}, on leaving the bubble the gas is swept down the side of the bubble and re-enters at the base. Thus an approximately spherical cloud of gas surrounds the bubble and moves with it up through the bed. In other words, there are in effect two phases formed within the bubble gas. This

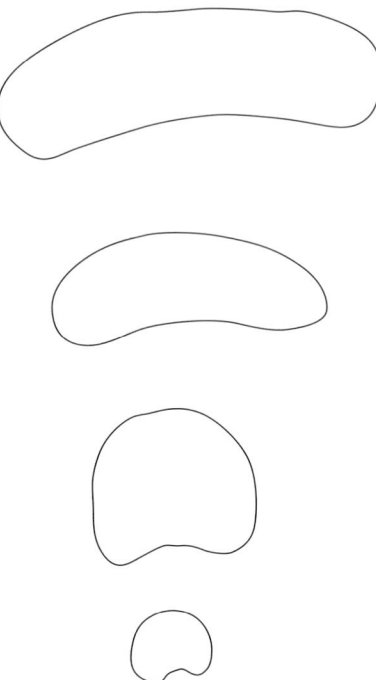

Figure 1.7   Characteristic bubble shapes.

phenomenon has significant consequences for the degree of gas interchange in aggregative fluidization and much effort has been expended
in modelling it in terms of the performance of fluidized bed chemical
reactors. As yet this appears to have little relevance to the food applications of fluidization although it may become significant in fluidized
bed fermentation when this operation is better understood.

Minimum bubbling velocity

Minimum bubbling velocity u_{mb} is defined as the gas velocity at which
bubbles first appear in aggregative fluidization. For coarse uniformly-
sized particles, for example those in Geldart group B, it is usually the
case that $u_{mb} \cong u_{mf}$. However, very fine non-uniformly sized particles
such as those in group A exhibit smooth bed expansion and no
bubbling until a gas velocity considerably in excess of the minimum
fluidizing velocity is reached. The ratio $\dfrac{u_{mb}}{u_{mf}}$, which indicates the
degree to which bed expansion occurs, is a strong function of mean
particle size and increases to values greater than unity, and perhaps
as high as 2 or 3, as the particle size falls below about $100\,\mu m$. According

to Abrahamson and Geldart (1980), this ratio depends particularly on the fraction of fines below $45\,\mu$m. They measured the minimum bubbling velocity for a wide range of fine particles in the range $20–72\,\mu$m and proposed the following correlation to predict $\dfrac{u_{mb}}{u_{mf}}$

$$\frac{u_{mb}}{u_{mf}} = \frac{2300\rho_f^{0.13}\mu^{0.52}\exp(0.72\omega_{45})}{d^{0.8}(\rho_s - \rho_f)^{0.93}} \qquad 1.5$$

where the mean particle size d, gas viscosity μ and the gas and solid densities ρ_f and ρ_s respectively are quoted in SI units and ω_{45} is the mass fraction of fines less than $45\,\mu$m in diameter.

Bubble rise velocity

The phenomenon of a gas bubble rising in a fluidized bed is similar to that of a gas bubble rising in a liquid; bubbles rise through the bed at a constant velocity for a given size, this velocity being proportional to bubble diameter. The general relationship for a single bubble is of the form

$$u_{SB} = k\sqrt{gd_B} \qquad 1.6$$

where, using SI units, it is usually assumed that k has a value of approximately 0.67 although for fine particles this underestimates velocity somewhat and Botterill (1975) suggests that k should be about 50% higher. Kunii and Levenspiel (1991) suggest $k = 0.71$ and that, based on a theory due to Davidson and Harrison (1963), the rise velocity for bubbles in a bubbling bed u_B is equal to the rise velocity of a single bubble plus the excess gas velocity, hence

$$u_B = u_{SB} + (u - u_{mf}) \qquad 1.7$$

To give an example of likely bubble rise velocities, take a fine powder with a mean particle diameter of $200\,\mu$m and a particle density of $1500\,\mathrm{kg\,m^{-3}}$. The Ergun equation (see equation 1.48) gives the minimum fluidizing velocity as $u_{mf} = 0.05\,\mathrm{m\,s^{-1}}$. Now calculating u_{SB}, the rise velocity of a single bubble, from equation 1.6 with $k = 0.71$, and the rise velocity in a bubbling bed from equation 1.7, gives the data in Table 1.1. Combinations of two fluidizing gas velocities and two bubble diameters have been assumed. This calculation suggests that in a bubbling bed of the given powder, bubbles are likely to rise at velocities in the approximate range $0.5–1\,\mathrm{m\,s^{-1}}$.

Table 1.1 Calculated rise velocity of bubbles.

Velocity (m s⁻¹)	$\dfrac{u}{u_{mf}} = 3$		$\dfrac{u}{u_{mf}} = 10$	
	$d_B = 0.025\,\text{m}$	$d_B = 0.075\,\text{m}$	$d_B = 0.025\,\text{m}$	$d_B = 0.075\,\text{m}$
u	0.15	0.15	0.50	0.50
$(u - u_{mf})$	0.10	0.10	0.45	0.45
u_{SB}	0.35	0.61	0.35	0.61
u_B	0.45	0.71	0.80	1.06

Particle movement due to bubble motion

Particle mixing in a fluidized bed is brought about solely by the movement of bubbles. The space between the indented base of a bubble and the bubble sphere is known as the wake and is occupied by particles. As a bubble rises in the bed (Figure 1.8) it carries with it this wake of particles and then draws up a spout of particles behind it. The wake grows as the bubble rises but a proportion of the wake may also be shed before the bubble reaches the bed surface. Thus the wake fraction, the fraction of the bubble sphere occupied by the wake, varies with time as a bubble rises to the bed surface. The wake fraction also varies with particle size in the approximate range 0.15–0.4 for a variety of (non-food) particles in the size range 60–600 μm; larger wake fractions have been observed with smaller particles but there is no clear relationship. Growth and shedding of the wake may be repeated several times in the life of a single bubble. Overall, a quantity of particles equal to one bubble volume is moved through a distance of 1.5 bubble diameters by a single rising bubble. Because of the large numbers of bubbles present in a fluidized bed, the particle mixing pattern is highly complex and extremely rapid.

Based upon these observations, Rowe (1977) derived an expression for the average particle circulation time t around a bed in terms of excess gas velocity and bed height at minimum fluidization

$$t = \frac{H_{mf}}{0.6(u - u_{mf})} \qquad\qquad 1.8$$

Although this expression may not accurately predict circulation time, and in any case particles do not follow a simple predetermined circuit around the bed, it serves to illustrate the significance of the excess gas velocity in determining particle mixing rates. The excess gas flow rate, proportional to the excess gas velocity, is essentially the bubble flow rate. A greater bubble flow generates more bubbles and therefore

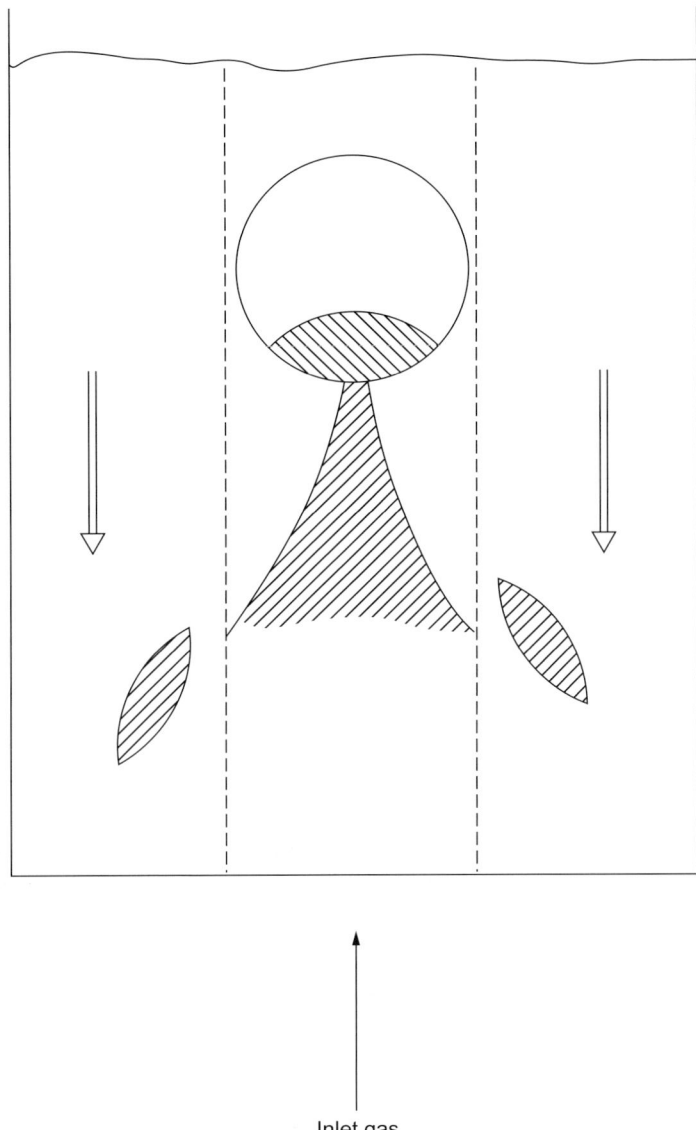

Inlet gas

Figure 1.8 Mechanism of particle mixing in a fluidized bed. From Smith, P.G., Introduction to food process engineering, Kluwer, 2003, figure 13.15. With kind permission of Springer Science and Business Media.

greater particle movement in unit time and more rapid particle circulation. The rapid particle movement in a fluidized bed leads to its use in particle mixing but is also essential in all other applications of fluidization. However, the mixing mechanism is opposed by segregation, where the concentration of one component increases at the bottom of the bed. Particle size has only a slight influence upon segregation (unlike in mechanical mixers). However, segregation increases markedly with the difference in density between two components. When the density ratio reaches a value of about 7, almost no mixing occurs in a fluidized bed. Three main mechanisms can be identified by which particles move in a bed and thus bring about either mixing or segregation:

- particles rising in bubble wakes
- large, dense particles falling through bubbles
- small, dense particles percolating downwards through the interstices of the dense phase of the bed.

Particle mixing is covered at greater length in Chapter 2.

Distributor plate design

A fluidized bed requires a distributor plate which supports the bed when it is not fluidized, prevents particles from passing through and promotes uniform fluidization by distributing the fluidizing medium evenly. In aggregative fluidization the nature of the distributor plate significantly influences the number and size of bubbles formed. A wide variety of distributors can be used including porous or sintered plates, manufactured from either ceramic or metal, layers of wire mesh, drilled plates, nozzles, tuyeres, bubble caps or pipe grids. Whatever the structure of the distributor plate, it must have many small orifices to allow gas to be injected at a multitude of points; a coarse distributor results in high gas rates at localised points and thus channelling in the bed (Richardson, 1971). Botterill (1975) summarised the requirements of a distributor which should:

(1) promote uniform and stable fluidization
(2) minimise the attrition of bed particles
(3) minimise erosion damage, and
(4) prevent the flow-back of bed material during normal operation and on interruption of fluidization when the bed is shut down.

Generally the pressure drop across the plate should be high to promote even gas distribution and is usually some fraction of the pressure drop across the fluidized bed; porous distributors tend to have much higher pressure drops than other types of plate.

Porous or sintered plates are the ideal and are used in small-scale studies of fluidized bed behaviour (Kunii and Levenspiel, 1991) and form a highly expanded unstable gas-solid dispersion directly above the distributor which rapidly divides into a large number of small bubbles plus an emulsion phase. Bubbles grow rapidly thereafter by coalescence. Kunii and Levenspiel (1991) also suggest that other materials offer similar properties to porous plates, including filter cloths, compressed fibres, compacted wire plates and even thin beds of small particles. According to Richardson (1971), porous plates are best in terms of the quality of fluidization but become expensive for large beds and also suffer from poor mechanical strength. Other disadvantages include the high pressure drop across the plate leading to increased power requirements and operating costs; sensitivity to thermal stresses; and blockage of the plate by fine particles or by the products of corrosion.

Perforated plate distributors are widely used in industry because they are cheap and relatively easy to manufacture. Simple perforated plate-type distributors suffer from particles passing back through to the plenum despite mean gas velocities well above the settling velocity for the particles. This is because of imbalances in gas flow between the orifices, which is difficult to eliminate. Hence, such plates take the form of either a layer of mesh sandwiched between two perforated plates or two staggered perforated plates without a mesh screen (Kunii and Levenspiel, 1991). However, these structures often lack rigidity and need to be reinforced or sometimes curved (concave to the bed) to withstand heavy loads. The diameter of the orifices in a perforated plate distributor varies from 1 or 2 mm in small beds used for research or very small-scale production to 50 mm in very large chemical reactors. Most food applications are likely to use apertures of intermediate size.

Perforated plate distributors cannot be used under severe operating conditions, such as high temperature or a highly reactive or corrosive environment. This is unlikely to be a disadvantage for food applications of fluidization, but in such circumstances tuyeres, nozzles or bubble caps are used. There is a very wide variety of designs (Kunii and Levenspiel, 1991) from open nozzles to complex bubble caps. The latter have small orifices around the periphery of a cap which rises or falls depending on the balance between the pressure of gas below and the back-pressure from above. The use of bubble caps can prevent the back-flow of solids (Botterill, 1975) and they may be designed to allow stagnant defluidized solids to lie between the caps, for example to provide thermal insulation for the protection of the distributor in high temperature processes. However, the potential disadvantages include the settling of particles between tuyeres or nozzles, the need to ensure that the incoming gas is free of fine particles which may clog the distributor, and the considerable expense of construction.

The pressure drop across the plate ΔP_d should be high to promote even gas distribution and stable fluidization and is usually some fraction of bed pressure drop ΔP although Kunii and Levenspiel (1991) point out that an excessive ΔP_d has the disadvantage of significantly increased power consumption and construction costs for the compressor. The pressure drop across a porous plate distributor increases in proportion to the superficial gas velocity whilst that for a perforated plate or a tuyere or nozzle type increases in proportion to the square of the gas velocity. It is important therefore to design for the minimum pressure drop that gives acceptable fluidization quality.

For a shallow bed, the pressure drop across the distributor should be of the same order as the bed pressure drop (Richardson, 1971). Hiby (1964) suggested $\dfrac{\Delta P_d}{\Delta P} = 0.15$ at low gas velocities of $\dfrac{u}{u_{mf}} \cong 1-2$

and $\dfrac{\Delta P_d}{\Delta P} = 0.015$ for $\dfrac{u}{u_{mf}} \gg 2$. This stability of fluidization with a relatively low plate pressure drop at higher gas velocities is because the fluidized bed behaves effectively as its own distributor and explains why ΔP_d can be a smaller fraction of ΔP with deeper beds. Siegel (1976) proposed a minimum value of 0.14 for $\dfrac{\Delta P_d}{\Delta P}$ whilst Shi and Fan (1984) suggested the same figure for porous plates and a value of 0.07 for perforated plates.

Characterisation of particulate solids

A knowledge of the minimum fluidizing velocity of fluidized solids is essential for the effective design and operation of fluidized beds which may be used for mixing, drying, freezing or other unit operations. The available predictive equations for minimum fluidizing velocity require a knowledge of the size, and sometimes the shape, of particles as well as the properties which characterise the particle bed. Some models require a knowledge of the terminal falling velocity of the particles. Therefore it is appropriate at this point to outline the ways in which particulate solids can be characterised and their bulk behaviour described, and in addition to give an elementary coverage of the interaction of a particle with a fluid before proceeding to the prediction of minimum fluidizing velocity.

Particle size distribution

Any sample of particulate food solids, whether naturally occurring or the result of a manufacturing process, will contain a distribution of particle sizes. However, despite the fact that the existence of a distribution of sizes can affect fluidized bed behaviour very significantly, predictive models, especially those which have been proposed for the prediction of minimum fluidizing velocity, usually require a single

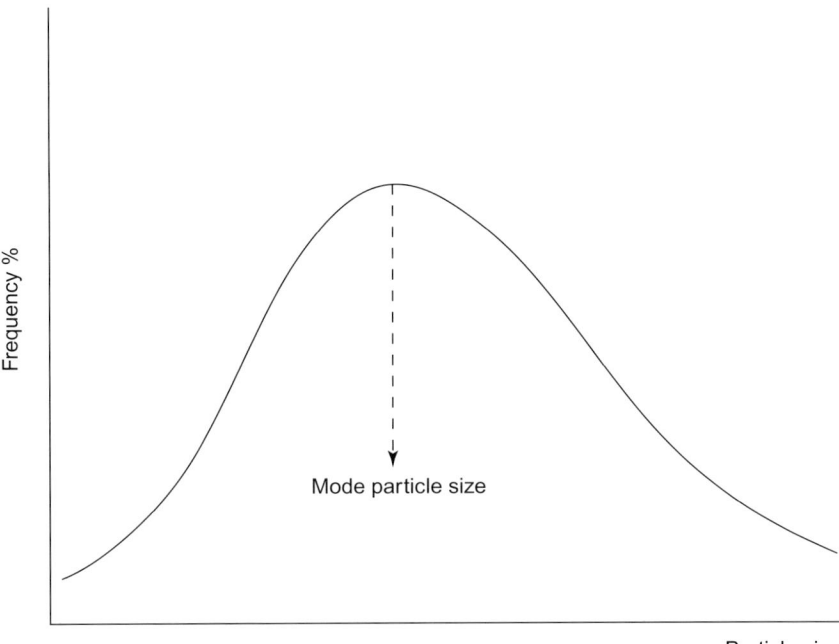

Figure 1.9 Frequency distribution curve.

value of particle size. Two expressions of 'average' size can be obtained from particle size distribution curves (Allen, 1981; Smith, 2003). The mode size is the most frequently occurring size and is represented by the peak of the frequency distribution curve, i.e. a plot of the frequency with which a given particle size occurs against that size (Figure 1.9). The frequency distribution curve also shows clearly the overall shape of a particle size distribution and the presence of very large or very small particles. Particle size data can also be plotted as a cumulative distribution in which either the cumulative percentage undersize or cumulative percentage oversize is plotted against size. These two curves are mirror images of each other (Figure 1.10) and the median particle size is the 50% point on either of the cumulative curves, which in turn must be the intersection of the undersize and oversize curves. In other words, it is that particle size which cuts the area under the frequency distribution curve in half. Although each of these quantities gives an indication of the average size, neither takes account of the spread of the distribution and it is possible to obtain the same mode or median with either a very narrow or a very wide distribution. These difficulties can be overcome by a using a carefully defined mean particle size.

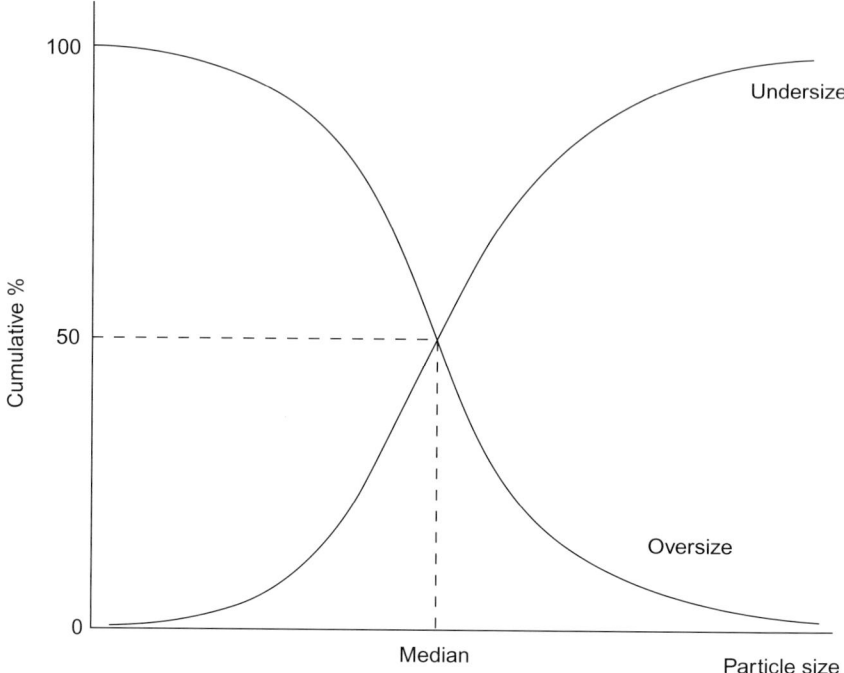

Figure 1.10   Cumulative percentage undersize and oversize curve.

Mean particle size

The characteristics of a particle size distribution can be defined as the total number, total length, total surface area and total volume of the particles. A distribution of particle sizes can be represented by a set of uniformly-sized particles which retains two characteristics of the original distribution. The mean particle size of a distribution is then equal to the size of the uniform particles with respect to the two characteristics. Thus, for the definition of mean particle size, Mugele and Evans (1951) proposed

$$x_{q,p}^{q-p} = \frac{\int\limits_{x_l}^{x_u} x^q \frac{dN}{dx} dx}{\int\limits_{x_l}^{x_u} x^p \frac{dN}{dx} dx} \qquad 1.9$$

where N is the number of particles of size x and p and q are parameters representing the characteristics of a distribution; x_l and x_u are the lower and upper limits of the size distribution respectively. Values of 0 for number, 1 for length, 2 for area and 3 for volume are assigned to the parameters p and q. Most methods of determining particle size generate data in the form of the number of particles within a given size band. The mean size is then obtained by summing over all the band widths and the model becomes

$$x_{q,p}^{q-p} = \frac{\sum(x^q N)}{\sum(x^p N)}$$

1.10

Substituting all the possible combinations of characteristics, i.e. values of p and q, into equation 1.10 gives rise to a number of different definitions of the mean size of a distribution. At minimum fluidization the drag force acting on a particle due to the flow of fluidizing gas over the particle is balanced by the net weight of the particle. The former is a function of surface area and the latter is proportional to particle volume. Consequently the surface-volume mean diameter, with $p = 2$ and $q = 3$, is the most appropriate particle size to use in expressions for minimum fluidizing velocity. It is defined by equation 1.11

$$x_{3,2} = \frac{\sum(x^3 N)}{\sum(x^2 N)}$$

1.11

Of the many ways of measuring or deducing the size of particles, sieving remains one of the easiest and cheapest and is very widely used. Although time consuming, it is the only method which gives a mass distribution and the only method which can be used for a reasonably large sample of particles which are not in suspension. A sieve analysis is carried out by placing a sample on the coarsest of a set of standard sieves made from woven wire. Below this sieve the other sieves are arranged in order of decreasing aperture size. The sample is then shaken for a fixed period of time, and the material on each sieve collected and weighed. For particularly fine or cohesive powders, air swept sieving can be used in which an upward flow of air from a rotating arm underneath the mesh prevents blockage of the sieve apertures.

The definition of the surface-volume mean diameter given by equation 1.11 must be modified for use with data from a sieve analysis. By assuming that the shape and density of the particles are constant for all size fractions, a number distribution can be transformed to a mass distribution (Smith, 2003) and therefore the surface-volume diameter becomes

$$x_{3,2} = \frac{1}{\sum\left(\dfrac{\omega}{x}\right)}$$

1.12

where ω is the mass fraction of particles of size x. The surface-volume mean is also known as the Sauter mean diameter or the harmonic mean diameter.

Particle shape

Relatively little appears to be known about the influence of shape on the behaviour of particulate solids and it is notoriously difficult to measure. Whilst a sphere may be characterised uniquely by its diameter and a cube by the length of a side, few natural or manufactured food particles are truly spherical or cubic. For irregular particles, or for regular but non-spherical particles, an equivalent spherical diameter d_e can be defined as the diameter of a sphere with the same volume V as the original particle. Thus

$$V = \frac{\pi}{6} d_e^3 \qquad\qquad 1.13$$

More commonly, a generalised volume shape factor K' is used to relate particle volume V to the cube of particle size

$$V = K'x^3 \qquad\qquad 1.14$$

The following has been suggested (Richardson and Zaki, 1954; Richardson, 1971) for fluidized cubes and cylinders

$$K'' = \frac{\pi}{6} \frac{d_s^3}{d_p^3} \qquad\qquad 1.15$$

where d_s is the diameter of a sphere with the same surface area as the particle and d_p is the diameter of a circle having an area equal to the projected area of the particle in its most stable position.

The sphericity ϕ of a particle, where the respective surface areas of the particle and an equivalent sphere are compared, has also been found to be useful in characterising shape. Thus

$$\phi = \frac{\text{surface area of sphere of equal volume to particle}}{\text{surface area of particle}} \qquad 1.16$$

For non-spherical particles, values of sphericity lie in the range $0 < \phi < 1$. Thus, the effective particle diameter for fluidization purposes is the product of the surface-volume mean diameter and the sphericity (Kunii and Levenspiel, 1991). The sphericity of regular-shaped particles can be deduced by geometry whilst the sphericity of irregular-shaped particles has to be determined by experiment. This is achieved by passing gas upwards through a packed bed of the particles at gas velocities below the minimum fluidizing velocity and measuring the packed bed interparticle voidage and frictional pressure drop. Kunii and Levenspiel (1991) suggest inserting these, and other appropriate values, into the Ergun equation (see equation 1.48) to obtain an average value for the sphericity. This procedure was used by Mishra *et al.* (1982) to estimate the sphericity for packed beds of grated yeast pellets.

Bulk particle properties

Expressions for minimum fluidizing velocity can be derived by examining the relationship between the velocity of a fluid passing through a packed bed of particles and the resultant pressure drop across the bed. Consequently it is necessary to define a number of bulk particle properties which influence fluidized bed behaviour.

The solids density ρ_s is the density of the solid material from which the particle is made and excludes any pore spaces within the particle. It can be measured using a specific gravity bottle and a liquid in which the particle does not dissolve. The envelope density of a particle is that which would be measured if an envelope covered the external particle surface, i.e. it is equal to the particle mass divided by the external volume. In most analyses the envelope and solids densities are assumed to be equivalent. The bulk density of a powder ρ_B is the effective density of the particle bed defined by

$$\rho_B = \frac{\text{mass of solids}}{\text{total bed volume}} \qquad 1.17$$

The bulk density will be considerably smaller than the solids density because the bed volume includes the volume of the spaces between particles.

Intraparticle porosity refers to the fraction of the particle volume which is occupied by internal pores; most manufactured food particles are porous. However, it is important to distinguish this quantity from bed voidage. The interparticle voidage ε is the fraction of the packed bed occupied by the void spaces between particles and is defined as

$$\varepsilon = \frac{\text{void volume}}{\text{total bed volume}} \qquad 1.18$$

This can be written as

$$\varepsilon = \frac{\text{total bed volume} - \text{particle volume}}{\text{total bed volume}} \qquad 1.19$$

or

$$\varepsilon = 1 - \frac{\text{particle volume}}{\text{total bed volume}} \qquad\qquad 1.20$$

Volume is inversely proportional to density and therefore equation 1.20 becomes

$$\varepsilon = 1 - \left(\frac{\rho_B}{\rho_s}\right) \qquad\qquad 1.21$$

The specific surface S is the external surface area of a particle per unit particle volume and for a sphere this is equal to $\dfrac{6}{d}$. The total surface area of a porous particle, including that of the internal pore spaces, can be measured by gas adsorption techniques and may be of the order of several hundred square metres per gram of material.

Terminal falling velocity and particle drag coefficient

Figure 1.11 represents the cross-section through a spherical particle over which an ideal non-viscous fluid flows. The fluid is at rest at points 1 and 3 but the fluid velocity is a maximum at points 2 and 4. There is a corresponding decrease in pressure from point 1 to point 2 and from 1 to 4. However, the pressure rises to a maximum again at point 3. If the ideal fluid is replaced with a real viscous fluid then, as the pressure increases towards point 3, the boundary layer next to the particle surface becomes thicker and then separates from the surface as in Figure 1.12. This separation of the boundary layer gives rise to

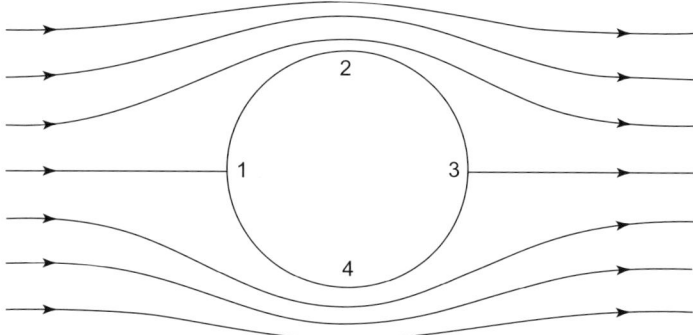

Figure 1.11 Fluid flow over a spherical particle. From Smith, P.G., Introduction to food process engineering, Kluwer, 2003, figure 13.3. With kind permission of Springer Science and Business Media.

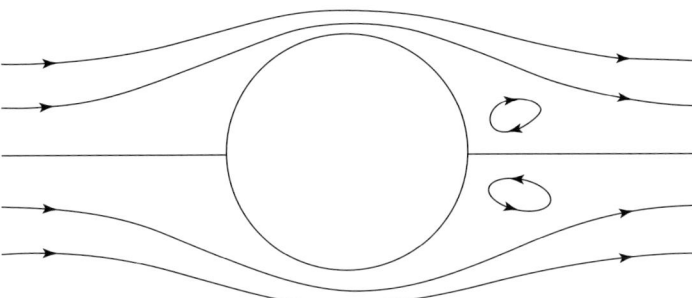

Figure 1.12 Separation of the boundary layer. From Smith, P.G., Introduction to food process engineering, Kluwer, 2003, figure 13.4. With kind permission of Springer Science and Business Media.

turbulent eddies within which energy is dissipated and which creates a force on the particle known as form drag. The total force acting on the particle because of the fluid flow is then the sum of form drag and the viscous drag over the surface.

A particle falling from rest, through a fluid, under gravity, will accelerate until it reaches a constant velocity known as the terminal falling velocity u_t. This velocity can be determined by balancing the product of particle mass and acceleration, the particle weight (mg) which pulls the particle down under gravity, the upthrust due to the fluid displaced as the particle falls ($m'g$) and the drag force (F) acting against the particle weight. Therefore

$$mg - m'g - F = m\frac{du}{dt} \qquad 1.22$$

At its terminal falling velocity the particle no longer accelerates and $\frac{du}{dt} = 0$. Substituting for this condition, for the mass (m) of a particle of diameter d and density ρ_s and for the mass of displaced fluid (m') of density ρ_f the drag force becomes

$$F = \frac{\pi}{6}g(\rho_s - \rho_f)d^3 \qquad 1.23$$

Stokes (1851) first showed that the drag force F on a sphere was given by

$$F = 3\pi d\mu u \qquad 1.24$$

where u is the relative velocity between the sphere and the fluid. Thus

$$3\pi d\mu u = \frac{\pi}{6}g(\rho_s - \rho_f)d^3 \qquad\qquad 1.25$$

and, putting $u = u_t$, the terminal falling velocity becomes

$$u_t = \frac{g(\rho_s - \rho_f)d^2}{18\mu} \qquad\qquad 1.26$$

This is Stokes' law which is valid in the particle Reynolds number range $10^{-4} < Re < 0.20$ where the Reynolds number is defined by

$$Re = \frac{\rho_f u_t d}{\mu} \qquad\qquad 1.27$$

and assumes that the particle is a single smooth, rigid sphere falling in a homogeneous fluid, that it is unaffected by the presence of any other particles and that the walls of the vessel do not exert a retarding effect on the particle.

A particle drag coefficient c_D can now be defined as the drag force divided by the product of the dynamic pressure acting on the particle (i.e. the velocity head expressed as an absolute pressure) and the cross-sectional area of the particle. This definition is analogous to that of a friction factor in conventional fluid flow. Hence

$$c_D = \frac{F}{\dfrac{\pi d^2}{4}\dfrac{\rho_f u^2}{2}} \qquad\qquad 1.28$$

On substituting for the drag force from equation 1.24, this gives

$$c_D = \frac{24}{Re} \qquad\qquad 1.29$$

for the Stokes region.

At Reynolds numbers beyond the Stokes region the boundary layer separates from the particle surface at a point just forward of the centre line of the sphere. A wake is formed containing vortices which results in larger frictional losses and a significantly increased drag force. This is known as the transition region $(0.20 < Re < 500)$ where no analytical solution for the drag force is possible and empirical equations must be used to describe the relationship between drag coefficient and Reynolds number. One of the most convenient and widely used of these is that due to Schiller and Naumann (1933)

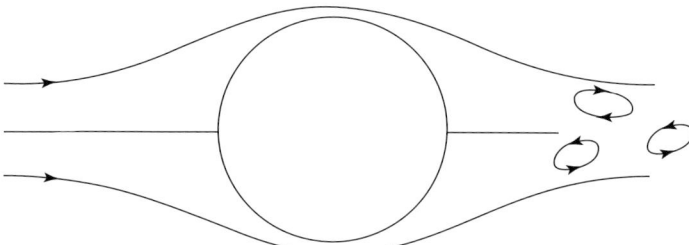

Figure 1.13 Vortex shedding. From Smith, P.G., Introduction to food process engineering, Kluwer, 2003, figure 13.4. With kind permission of Springer Science and Business Media.

$$c_D = \frac{24}{Re}(1 + 0.15Re^{0.687}) \qquad\qquad 1.30$$

As the Reynolds number increases further, vortex shedding takes place (Figure 1.13) in what is known as the Newton region, in the range $500 < Re < 2 \times 10^5$, where the drag coefficient has a value of approximately 0.44. Consequently equation 1.28 gives the drag force as

$$F = 0.055\pi d^2 \rho_f u^2 \qquad\qquad 1.31$$

which, on substitution into equation 1.23, gives the terminal falling velocity as

$$u_t = 1.74\sqrt{\frac{dg(\rho_s - \rho_f)}{\rho_f}} \qquad\qquad 1.32$$

At still greater Reynolds numbers the boundary layer itself becomes turbulent and separation occurs at the rear of the sphere and closer to the particle. In this fully turbulent region, beyond $Re = 2 \times 10^5$, the drag coefficient falls further to a value of about 0.10.

Minimum fluidizing velocity in aggregative fluidization

The superficial velocity of gas in a fluidized bed, relative to the minimum fluidizing velocity, is the quantity which has the greatest influence on the behaviour of a given particle bed. Consequently, a knowledge of minimum fluidizing velocity is vital to the operation of fluidized beds and much research effort has been expended in attempting to predict it.

Voidage and pressure drop at incipient fluidization

At the point of incipient fluidization the drag force exerted on a particle is equal to its net weight. For the whole particle bed the drag force can be equated to the product of bed pressure drop ΔP and bed cross-sectional area A. The net bed weight is then the product of bed volume, net density, the fraction of the bed $(1 - \varepsilon)$ which is occupied by particles and the acceleration due to gravity. Thus, at minimum fluidizing velocity

$$\Delta P_{mf} A = A H_{mf} (\rho_s - \rho_f)(1 - \varepsilon_{mf})g \qquad 1.33$$

or, eliminating the cross-sectional area

$$\Delta P_{mf} = (\rho_s - \rho_f)(1 - \varepsilon_{mf})g H_{mf} \qquad 1.34$$

Alternatively equation 1.34 may be thought of as equating the hydrostatic pressure at the base of a column of fluid to the product of bed height, density and the acceleration due to gravity. However, in the case of fluidized solids the density is equal to the difference in density between the particle and the fluidizing medium, the term $(1 - \varepsilon)$ being included because it is only the particles which contribute significantly to the pressure drop. Equation 1.34 allows bed voidage to be determined from experimental measurements of bed pressure drop and bed height. In practice, bed voidage is a function of particle size, particle shape and particle size distribution. Richardson (1971) suggests an approximate value of $\varepsilon_{mf} = 0.4$ for spherical particles. This is perhaps a little low, with values being closer to 0.5 for many particles (Leva, 1959; Kunii and Levenspiel, 1991).

Carman-Kozeny equation

Minimum fluidizing velocity can be predicted from a knowledge of the relationship between the velocity of a fluid passing through a bed of particles and the consequent pressure drop across the bed. However, modelling this relationship is inherently difficult because of the irregular nature of the void spaces in a bed of irregular non-uniformly sized particles; an exact solution is not possible because of the tortuous flow paths followed by the fluid (Figure 1.14). The Carman-Kozeny model is based upon the idea that a packed bed can be modelled by a series of capillaries (Figure 1.15), to which the Hagan-Poiseuille relationship for laminar flow in a tube is applied. By using an equivalent diameter d' and length L' to represent the void spaces, the interstitial velocity u' (i.e. the velocity between the particles in a packed bed) is given by

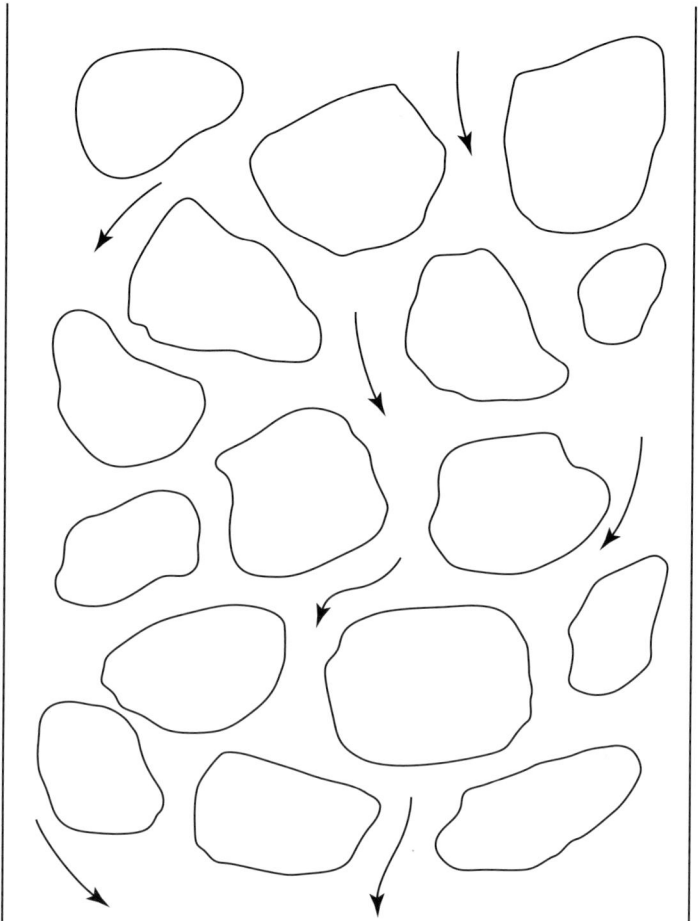

Figure 1.14 Flow path through a packed bed of irregular non-uniformly sized particles.

$$u' = \frac{\Delta P d'^2}{32 L' \mu}$$ 1.35

If now the cross-sectional area which determines the interstitial velocity for a given volumetric flow rate is proportional to the interparticle voidage, it follows that

$$u' = \frac{u}{\varepsilon}$$ 1.36

Kozeny further suggested that the equivalent pore space diameter d' is given by

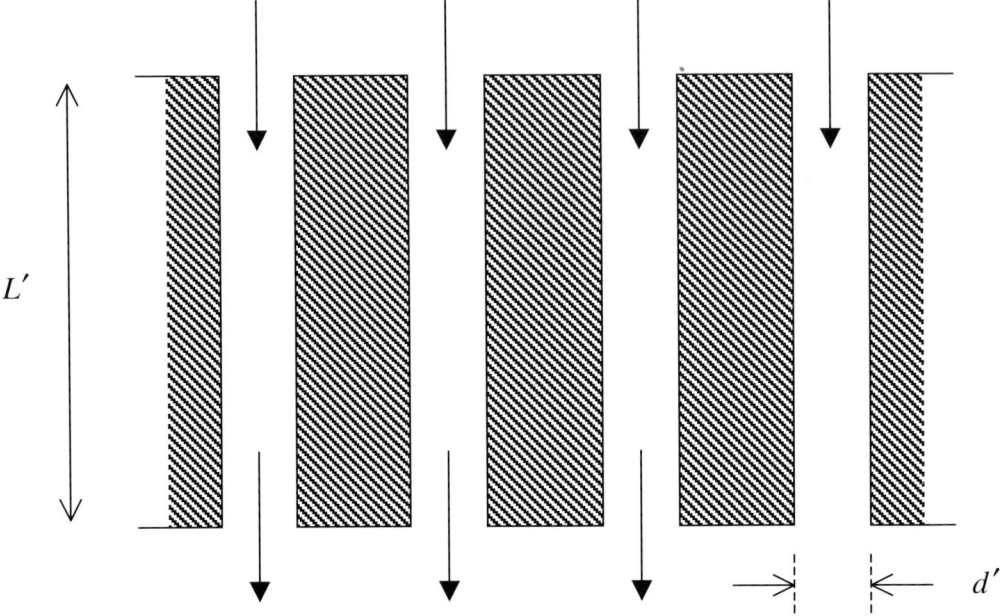

Figure 1.15   Carman-Kozeny model of a packed bed.

$$d' = \frac{\varepsilon}{S_B}$$
1.37

where S_B is the particle surface area per unit bed volume which comes into contact with the fluid passing through the bed. In turn, this quantity is related to the specific surface by the fraction of the bed occupied by particles $(1 - \varepsilon)$. Thus

$$S_B = S(1 - \varepsilon)$$
1.38

Substituting each of these assumptions into equation 1.35, and further assuming that the equivalent pore space length is proportional to the bed depth H, results in the Carman-Kozeny equation

$$u = \frac{\varepsilon^3 \Delta P}{KS^2(1 - \varepsilon)^2 \mu H}$$
1.39

which relates superficial fluid velocity and pressure drop for a bed of incompressible particles, i.e. for cases where the bed voidage is constant.

The Carman-Kozeny expression for minimum fluidizing velocity is obtained by substituting for the pressure drop at minimum fluidization from equation 1.34 and therefore

$$u_{mf} = \frac{\varepsilon_{mf}^3 (\rho_s - \rho_f) g}{KS^2 (1 - \varepsilon_{mf}) \mu}$$ 1.40

The dimensionless constant K is known as Kozeny's constant and has a value of approximately 5.0 although strictly it is a function of both intraparticle porosity and particle shape. Assuming now that the bed particles are spherical for which the specific surface is equal to $\frac{6}{d}$, and that $K = 5$, the minimum fluidizing velocity is given by

$$u_{mf} = \frac{\varepsilon_{mf}^3 (\rho_s - \rho_f) g d^2}{180 (1 - \varepsilon_{mf}) \mu}$$ 1.41

This relationship is based on an assumption of laminar flow between particles in a packed bed. Consequently the particle Reynolds number is limited to $Re_{mf} < 20$ (Kunii and Levenspiel, 1991) where

$$Re_{mf} = \frac{\rho_f u_{mf} d}{\mu}$$ 1.42

This corresponds to relatively fine particles, certainly below $500\,\mu$m in diameter and perhaps smaller, and thus can be used only for fine food powders, perhaps those encountered in drying and mixing operations, and not for the larger particulates such as vegetable pieces in fluidized bed freezers or large agglomerates in granulation systems. Note that equation 1.41 suggests that u_{mf} is proportional to the difference in density between particle and fluid, proportional to the square of particle diameter and inversely proportional to fluid viscosity.

Ergun equation

The Carman-Kozeny equation works well for fine particles. However, for large particles, greater than about 600 or $700\,\mu$m, the Carman-Kozeny relationship is inadequate and predicts far too low a pressure drop. For larger particles, for example peas in a fluidized bed freezer, the minimum fluidizing velocity is high and the kinetic energy losses are significant. In these circumstances the Carman-Kozeny equation vastly overestimates u_{mf} and the Ergun equation is more accurate. This is a semi-empirical equation for the pressure drop per unit bed depth, containing two terms. Ergun's equation may be expressed as

$$\frac{\Delta P}{H} = \frac{150 (1 - \varepsilon)^2 \mu u}{\varepsilon^3 \phi^2 d^2} + \frac{1.75 (1 - \varepsilon) \rho_f u^2}{\varepsilon^3 \phi d}$$ 1.43

The first term represents the pressure loss due to viscous drag (this is essentially the Carman-Kozeny equation) whilst the second term represents kinetic energy losses, which are significant at higher velocities (kinetic energy being proportional to velocity squared). Equation 1.43 is valid in the range $1 < Re < 2000$ where the Reynolds number is defined by

$$Re = \frac{u\rho}{S(1-\varepsilon)\mu}$$ 1.44

Writing Ergun's equation for minimum fluidizing conditions, and assuming the particles to be spherical ($\phi = 1$), gives

$$\frac{\Delta P}{H_{mf}} = \frac{150(1-\varepsilon_{mf})^2 \mu u_{mf}}{\varepsilon_{mf}^3 d^2} + \frac{1.75(1-\varepsilon_{mf})\rho_f u_{mf}^2}{\varepsilon_{mf}^3 d}$$ 1.45

and substituting for pressure drop from equation 1.34 gives

$$(\rho_s - \rho_f)g = \frac{150(1-\varepsilon_{mf})\mu u_{mf}}{\varepsilon_{mf}^3 d^2} + \frac{1.75\rho_f u_{mf}^2}{\varepsilon_{mf}^3 d}$$ 1.46

This expression is rather unwieldy but can be simplified considerably. Multiplying through by $\dfrac{\rho_f d^3}{\mu^2}$ results in

$$\frac{\rho_f(\rho_s - \rho_f)d^3 g}{\mu^2} = \frac{150(1-\varepsilon_{mf})u_{mf}d\rho_f}{\varepsilon_{mf}^3 \mu} + \frac{1.75\rho_f^2 d^2 u_{mf}^2}{\varepsilon_{mf}^3 \mu^2}$$ 1.47

which can be put into the form

$$Ga = \frac{150(1-\varepsilon_{mf})}{\varepsilon_{mf}^3} Re_{mf} + \frac{1.75 Re_{mf}^2}{\varepsilon_{mf}^3}$$ 1.48

This is a quadratic equation in Re_{mf} where the Galileo number Ga is defined by

$$Ga = \frac{\rho_f(\rho_s - \rho_f)d^3 g}{\mu^2}$$ 1.49

and the particle Reynolds number at minimum fluidization by equation 1.42. The Galileo number is also known as the Archimedes number which is usually given the symbol Ar.

Knowledge of the size and density of the particles to be fluidized, of the density and viscosity of the fluidizing gas (at the relevant temperature) as well as a knowledge of the interparticle voidage at minimum fluidizing conditions (which can be obtained from equation 1.34) allows the Ergun equation to be solved for Re_{mf}. This in turn will permit u_{mf} to be found with the already known values of d, μ and ρ_f. For very large particles the first (viscous drag) term is dominated by the kinetic energy term and it can safely be ignored. Equation 1.48 then reduces to

$$Re_{mf}^2 = \frac{\varepsilon_{mf}^3 Ga}{1.75}$$
1.50

a slightly more convenient relationship. Kunii and Levenspiel (1991) suggest that such a procedure is valid for $Re_{mf} > 1000$. Smith (2003) and Jackson and Lamb (1981) give worked examples illustrating how both the Ergun and Carman-Kozeny equations can be used to determine the minimum fluidizing velocity of food particulates. Smith demonstrates that for particles with diameters of several millimetres (such as peas or small vegetable pieces), the Carman-Kozeny equation vastly overestimates minimum fluidizing velocity. Further, for such particles, the error in leaving out the viscous drag term in the Ergun equation is small and of the order of 2%.

Minimum fluidizing velocity as a function of terminal falling velocity

Richardson (1971) summarises a method of predicting minimum fluidizing velocity as a function of the terminal falling velocity of a particle. This requires the terminal falling velocity u_t to be expressed in terms of the Galileo number. Thus, treating the Stokes, transition and Newton regions in turn:

(1) For the Stokes region, if equation 1.26 is multiplied by $\dfrac{\rho_f d}{\mu}$ the result is

$$\frac{\rho_f u_t d}{\mu} = \frac{g(\rho_s - \rho_f)d^3 \rho_f}{18\mu^2}$$
1.51

which in terms of the Galileo number (equation 1.49) is

$$Re_t = \frac{Ga}{18}$$
1.52

or

$$Ga = 18 Re_t$$
1.53

Stokes' law is valid for Reynolds numbers below 0.20 which becomes, for equation 1.53, $Ga < 3.6$.

(2) For the transition region, substituting for the definitions of drag coefficient (equation 1.28) and drag force (equation 1.24) in the Schiller and Naumann equation for drag coefficient gives

$$\frac{\pi}{6}g(\rho_s - \rho_f)d^3 = \frac{\pi d^2}{4}\frac{\rho_f u^2}{2}\frac{24}{Re_t}(1 + 0.15Re_t^{0.687}) \qquad 1.54$$

When multiplied by $\dfrac{6\rho_f}{\mu^2}$ this results in

$$\frac{\rho_f g(\rho_s - \rho_f)d^3}{\mu^2} = \frac{18\rho_f^2 d^2 u^2}{\mu^2}\frac{1}{Re_t}(1 + 0.15Re_t^{0.687}) \qquad 1.55$$

which in turn becomes

$$Ga = 18Re_t(1 + 0.15Re_t^{0.687}) \qquad 1.56$$

and which is valid in the range $3.6 < Ga < 10^5$.

(3) Finally, for the Newton region, at Reynolds numbers greater than 2×10^5, squaring equation 1.32 and multiplying by $\dfrac{\rho_f^2 d^2}{\mu^2}$ yields

$$\frac{u_t^2 \rho_f^2 d^2}{\mu^2} = (1.74)^2 \frac{\rho_f^2 d^2}{\mu^2}\frac{dg(\rho_s - \rho_f)}{\rho_f} \qquad 1.57$$

In terms of the Reynolds number this becomes

$$Re_t^2 = (1.74)^2 \frac{\rho_f d^3 g(\rho_s - \rho_f)}{\mu^2} \qquad 1.58$$

and hence

$$Ga = 0.33Re_t^2 \qquad 1.59$$

for Galileo numbers greater than about 10^5. If now the voidage at minimum fluidizing conditions ε_{mf} is known, then for a given value of the Galileo number Ga the ratio of terminal falling velocity to minimum fluidizing velocity $\left(\dfrac{u_t}{u_{mf}}\right) = \left(\dfrac{Re_t}{Re_{mf}}\right)$ can be calculated from the Ergun equation and one of equations 1.53, 1.56 or 1.59 respectively. Richardson presents a plot of $\left(\dfrac{u_t}{u_{mf}}\right) = \left(\dfrac{Re_t}{Re_{mf}}\right)$ against Ga for commonly

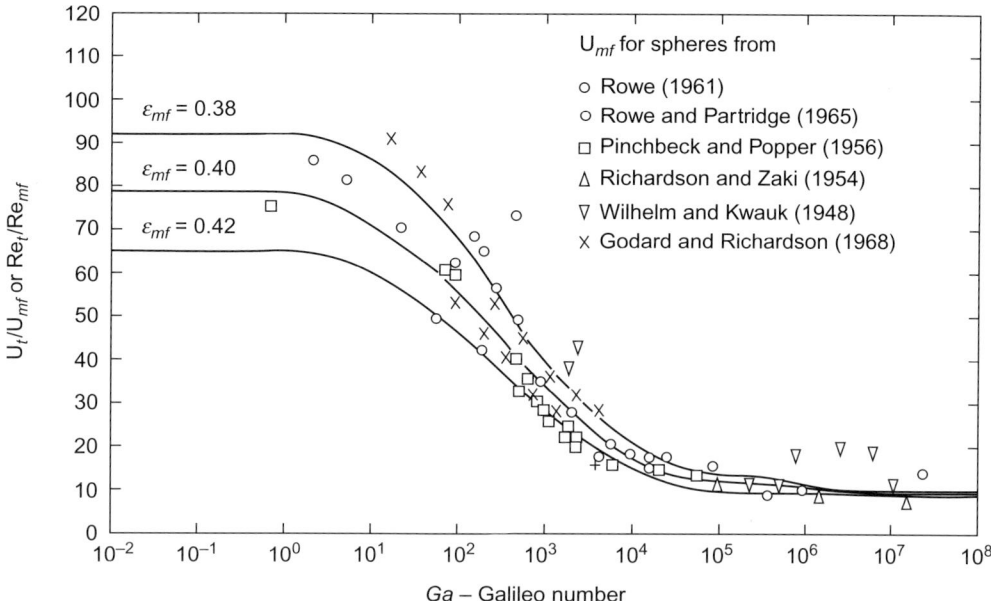

Figure 1.16   Ratio of terminal falling velocity to minimum fluidizing velocity as a function of Galileo number. Reprinted from Davidson, J.F. and Harrison, D., Fluidization, Academic Press, 1971, with permission from Elsevier.

encountered values of voidage (i.e. 0.38, 0.40 and 0.42 respectively) (Figure 1.16). Thus a knowledge of fluid density and viscosity, particle density and diameter allows the calculation of Ga and thus u_{mf}. The values of $\left(\dfrac{u_t}{u_{mf}}\right)$ are constant for $Ga < 3.6$ and for $Ga > 10^5$; the procedure is rather less convenient in the transition region.

Semi-empirical correlations

Probably the most useful and accurate of the many semi-empirical equations available is that due to Leva

$$u_{mf} = 0.0079 \frac{d^{1.82}(\rho_s - \rho_f)^{0.94}}{\mu^{0.88}} \qquad 1.60$$

This equation allows the prediction of minimum fluidizing velocity from a knowledge of the mean particle diameter, the particle density, the density of fluidizing medium and the viscosity of fluidizing medium (SI units). Couderc (1985) quotes data which show that the inaccuracy of Leva's equation increases significantly outside the range $2 < Re < 30$.

A number of empirical variations on the Ergun equation have been proposed. It can be seen that equation 1.48 is of the form

$$\alpha Re_{mf}^2 + \beta Re_{mf} - Ga = 0 \qquad\qquad 1.61$$

This can be rearranged to give

$$Re_{mf} = \left[\left(\frac{\beta}{2\alpha} \right)^2 + \left(\frac{Ga}{\alpha} \right) \right]^{0.5} - \left(\frac{\beta}{2\alpha} \right) \qquad\qquad 1.62$$

Recognising that α and β change very little over a very wide range of conditions, a number of workers have fitted experimental data for minimum fluidizing velocity to the Ergun form (Kunii and Levenspiel, 1991). The resulting correlations require neither a knowledge of ε_{mf} nor of ϕ in order for them to be used to predict u_{mf}. Two examples widely used are those due to Wen and Yu (1966) and to Saxena and Vogel (1977), equations 1.63 and 1.64 respectively

$$Re_{mf} = [(33.7)^2 + (0.0408\,Ga)]^{0.5} - 33.7 \qquad\qquad 1.63$$

$$Re_{mf} = [(25.3)^2 + (0.0571Ga)]^{0.5} - 25.3 \qquad\qquad 1.64$$

Experimental measurement

The standardised procedure for measuring minimum fluidizing velocity is to fluidize a bed of particles vigorously for some minutes and then reduce gas velocity in small increments to overcome the hysteresis arising from frictional forces in the bed, recording the bed pressure drop each time. This may be done with a simple water manometer with one leg open to atmosphere and one leg connected to a narrow tube placed in the bed and just below the bed surface. The data are then interpreted as in Figure 1.3; u_{mf} corresponds to the intersection of the straight lines representing the fixed and fluidized beds.

Fluidized bed behaviour at high gas velocities

Slugging

As the superficial gas velocity increases, the nature of the bubbles changes. Especially in beds of small diameter or in deep beds, i.e. those with a bed depth to diameter ratio greater than unity, and with fine particles, the bubbles grow to the size of the bed container and push plugs of material up the bed as they rise. The particles then stream past

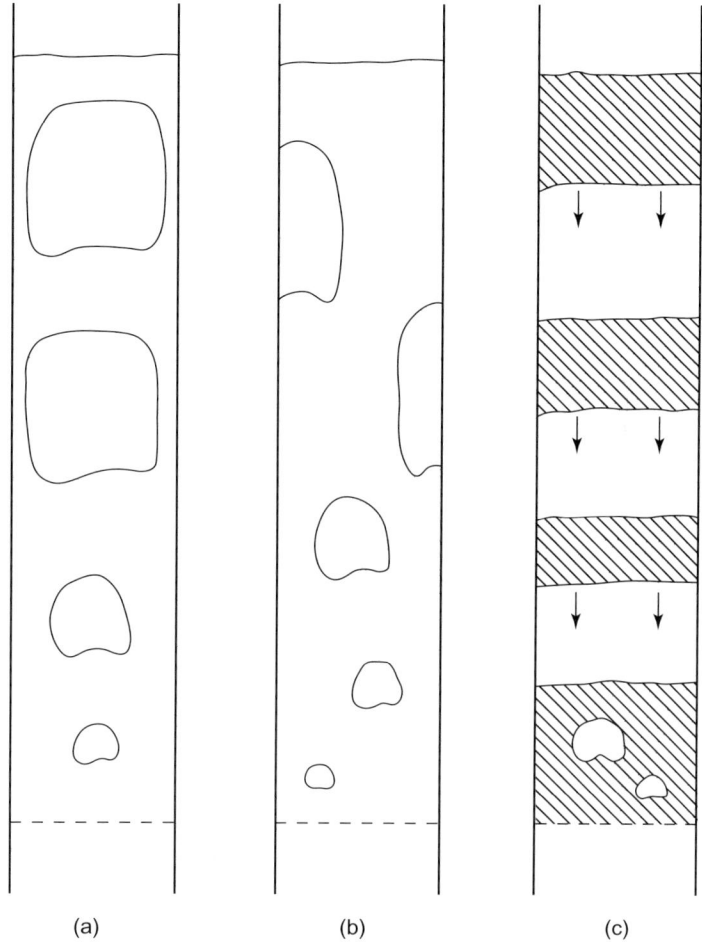

Figure 1.17   Types of slugging behaviour.

the slugs of gas at the bed walls on their downward path as shown in Figure 1.17a. This is known as axial slugging and generally is to be avoided in food processing applications, although it does find application when a fluidized bed is used as a chemical reactor. A single slug tends to rise as would a large bubble in a liquid. Thus the rise velocity is given by a relationship of the form of equation 1.6 but with $k = 0.35$ and is a function of the bed diameter rather than the bubble diameter.

In a second type of slugging behaviour the slugs of gas cling to the wall of the bed as in Figure 1.17b; this tends to occur either with angular particles or when the vessel wall is rough. In the third type of slugging, Figure 1.17c, which occurs with group D particles, there are no distinct bubbles but instead the bed is divided into alternate regions of dense phase and disperse phase with approximately horizontal interfaces. Particles rain down from one slice of dense phase through a region of disperse phase to the next slice of dense phase below.

Baeyens and Geldart (1974), in an experimental study using a wide range of particle sizes and bed diameters, suggested that generally slugging does not occur below a bed height of H_s given by

$$H_S = 0.6D_{bed}^{0.175}$$

1.65

where D_{bed} is the bed diameter.

Turbulent fluidization and fast fluidization

As the superficial gas velocity is increased still further the two-phase bed structure, of a bubble or slug phase plus a dense phase, begins to break down and the bed is characterised by significant pressure fluctuations. The bed is now referred to as a turbulent bed where discrete bubbles or slugs are no longer present. There is far more particle ejection into the freeboard and a far less distinct interface between the dense-phase bed and the freeboard. The upper limit of turbulent fluidization is the point where particles are transported out of the column. Elutriated or transported solids must now be recycled to the bed via a cyclone if any solids are to be retained in the fluidized column. If the return feed rate to the bed is low then the solid concentration which results in the bed gives rise to a dilute phase of low concentration which is similar to that found in pneumatic conveying. However, if the solids feed rate is high then a higher concentration of solids can be maintained in the bed and this is known as fast fluidization; gas velocities are very high, at least an order of magnitude greater than the terminal falling velocity. There is an extensive literature on so-called high velocity fluidization, which includes both turbulent and fast fluidization. However, as with slugging, there are few if any applications involving food materials and these regimes are relevant only to heterogeneous chemical reactors.

Elutriation and entrainment

In any fluidized bed operating with a gas velocity above u_{mf} some particles will be transported into the gas stream above the bed surface; this is called entrainment. Some of these particles will be transported sufficiently so as to leave the fluidized bed column entirely and this mass of solids is then referred to as carry-over. However, the phenomenon is a little more complex. Fractionation or preferential separation of the bed particles occurs and this changes with height above the bed surface; it is this phenomenon which is called elutriation. In a bed with a distribution of particle sizes, and most fluidized beds of real interest come into this category, only the smallest particles are removed and form a disperse phase which occupies the freeboard. The freeboard is the term given to the space between the surface of the dense phase and the point at which gas exits the column. It is important to differentiate, however, between the disperse phase and so-called dilute-phase fluidization where all the particulate solids are carried in the gas stream and which corresponds to one of the regimes of pneumatic conveying.

Elutriation is important in most industrial fluidized beds and is generally thought of as a disadvantage. In addition to the small particles which may be present in the initial particle size distribution, fines may be created in the course of operation by the attrition of bed particles. Elutriated particles usually need to be collected and recovered either because they represent the loss of product particles of a given size, because they must be separated from the exhaust gas for environmental reasons, or because of safety concerns; there is a considerable risk of a dust explosion with very fine particles and perhaps especially so with many food particulates. Therefore the fluidized bed plant will require ancillary gas cleaning equipment such as a cyclone, filter or electrostatic precipitator to separate the fines from the gas. The loss of a particular size fraction from the bed may change fluidized bed behaviour and it then becomes important to return the fines to the bed continuously.

Elutriation can occur only if the terminal falling velocity of a particle is exceeded. Particles are brought to the surface of the bed by the rising gas bubbles which carry particles up the bed in the bubble wake and the eruption of the bubbles at the dense phase surface then throws particles up into the freeboard where the local velocity is at least approximately equal to the superficial velocity and possibly higher. In addition, gas bubbles rise through the bed at a velocity higher than that of the surrounding dense phase and also the pressure in the bubble is very slightly higher than the surrounding interstitial gas. These factors explain why the particles in the bubble wake material are thrown into the freeboard as packets of particles. The fines are then carried up in the gas stream and the larger particles fall back.

There is, however, considerable variation across a bed diameter in the velocity of the gas leaving the surface of the bed simply because any given bed cross-section contains both bubble phase and lean phase (Figure 1.18). As the gas travels up the column these variations in velocity become less pronounced and therefore the larger particles fall back as the local velocity falls below their terminal falling velocity. The height above the bed surface at which the larger particles disengage is

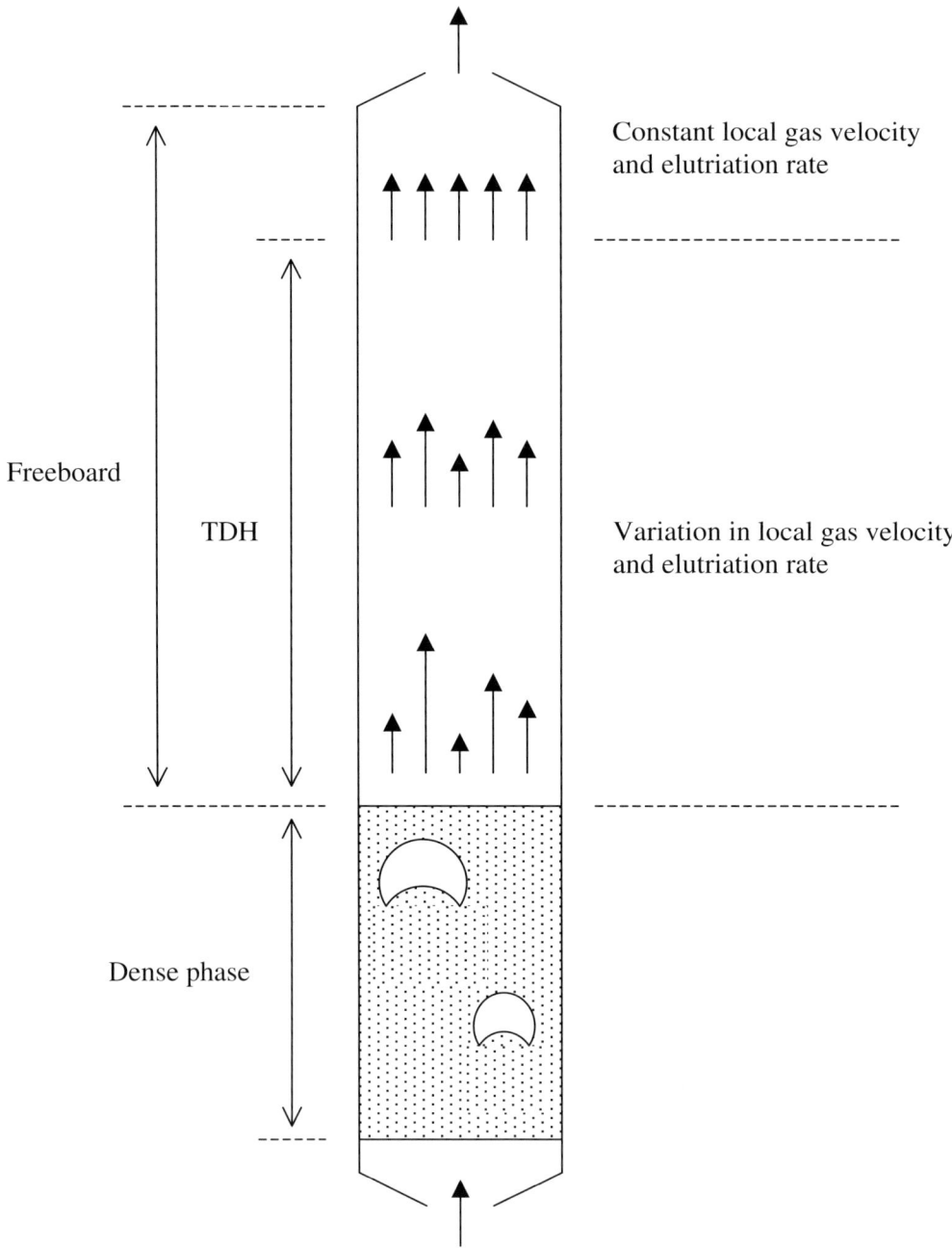

Figure 1.18   Elutriation and transport disengaging height (TDH).

called the transport disengaging height (TDH). Alternatively the TDH may be defined as the height at which the elutriation rate, and therefore the particle size distribution of the elutriated particles, is approximately constant. The TDH increases with both superficial gas velocity and with bed diameter; the latter effect is due to wall effects at small bed diameters and to relatively poor gas distribution at larger bed diameters. The concentration of solids decreases with height above the bed surface and therefore increasing the height of the freeboard reduces the loss of particles from the bed. In other words, when the gas exit is above the TDH the rate of solids loss is constant and therefore it is common to design the freeboard height to be equal to TDH.

Other types of fluidization

Spouted beds

A spouted bed requires a cylindrical container with an inverted conical base and a vertical pipe entering at the apex of the cone. If coarse solids are placed in the container and gas is introduced through the pipe, the solids will begin to 'spout' as in Figure 1.19. Particles are carried at high velocity upwards in the central jet in a dilute phase and at the bed surface the particles form a kind of fountain and rain down onto the clearly defined surface. They then travel down the bed in the annulus at much lower velocities than those in the central spout. On reaching the gas inlet point, the particles are re-entrained and thus complete a well-defined and regular cycle. Spouted beds can exist only with large particles, usually greater than 1–2 mm in diameter. A uniform particle size distribution is more likely to result in spouting bed behaviour, although particles with a larger mean diameter will spout more readily with a wider size distribution than will smaller particles. If the cone is too steep then the bed becomes unstable; the limiting cone angle is about 40°. Particle motion is much more regular than in a fluidized bed although there is a distribution of cycle times with particles re-entering the spout at various points up the bed.

As in the case of fluidization, there is a gas velocity below which spouting does not occur. The earliest and best known relationship to predict the minimum spouting velocity u_{ms} is that due to Mathur and Gishler (1955)

$$u_{ms} = \left(\frac{d}{D_{bed}}\right)\left(\frac{D_i}{D_{bed}}\right)^{0.33}\left(\frac{2gH(\rho_s - \rho_f)}{\rho_f}\right)^{0.5} \qquad 1.66$$

which indicates that the minimum spouting velocity increases with bed depth H but decreases with bed diameter D_{bed} and that it increases

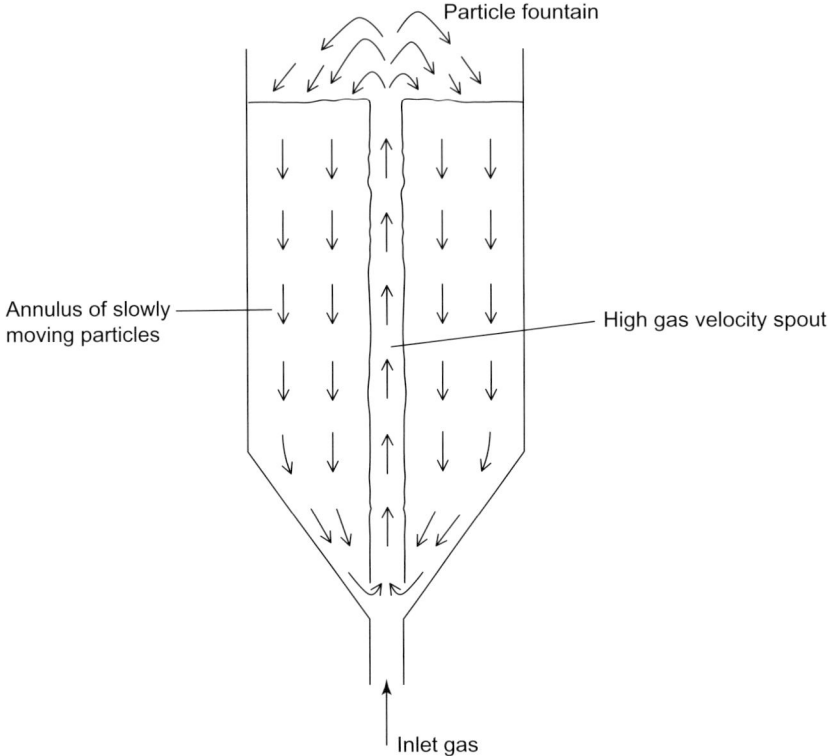

Figure 1.19 Spouted bed.

slightly with the diameter of the gas inlet D_i. The pressure drop down
the bed is not uniform; it is small at the base and reaches a maximum
at the surface where it approaches the pressure drop needed to support
the solids. If the gas velocity is sufficient to fluidize the particles then
the spouted bed becomes unstable; thus there is a maximum spoutable
depth H_{sm} beyond which spouting behaviour breaks down and the bed
becomes fluidized. Mathur (1971) suggests that u_{ms} is approximately
equal to the minimum fluidizing velocity when the bed depth is close
to the maximum spoutable depth. However, because in practical situ-
ations the bed depth would be considerably less than this maximum,
gas velocities required for spouting are somewhat less than those
required for fluidization. Data are available (Mathur, 1971) for a variety
of solids, including wheat, mustard seed, rape seed, millet and peas,
for all of which equation 1.66 has been shown to be valid. The simplest
way to determine H_{sm} is to calculate, or measure, the minimum fluidizing velocity of the bed particles and then substitute this figure for
u_{ms} in equation 1.66, solving for H. Predicted values of maximum
spoutable depth obtained with this approach are in agreement with
experimental data (Mathur, 1971). The applications of spouted beds

include drying, particularly the drying of wheat, granulation and the coating of seeds and large particles.

Centrifugal fluidization

The so-called centrifugal fluidized bed has found application in the drying of foods and is claimed to give stable and smooth fluidization of large irregular particles at velocities well above those required for pneumatic transport. Its particular advantage is that high gas velocities allow significant heat input to the bed at lower temperatures than would otherwise be needed and hence the high rates of heat transfer which become possible in the constant drying rate zone of a drying process allow the centrifugal bed to be used for the reduction of initial moisture in sticky, high-moisture, heat-sensitive foods (Carlson *et al.*, 1976).

Hanni *et al.* (1976) described a continuous fluidized bed drier using the centrifugal principle. It consisted of a perforated horizontal cylinder rotating inside a plenum with hot air blown across the outside of the cylinder and perpendicular to the axis of rotation (Figure 1.20). Vanes placed in the air inlet allowed the incident angle of the air flow to be varied from 0° to 45° to the perpendicular. Particles were fluidized inside the Teflon-coated stainless steel cylinder which could be tilted by up to 6° in order to control particle residence time. The cylinder, of

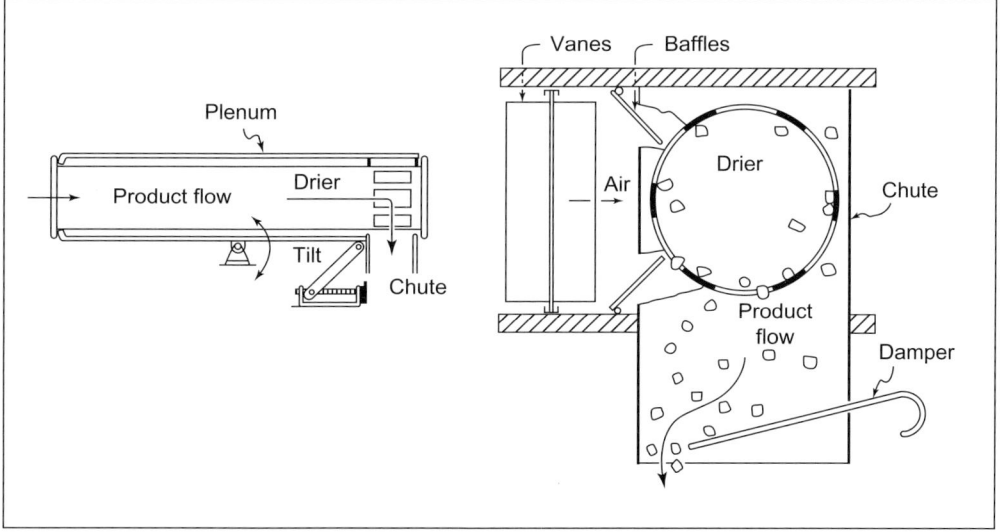

Figure 1.20 Centrifugal fluidized bed. From Hanni *et al.* (1976), by permission of the Institute of Food Technologists, USA.

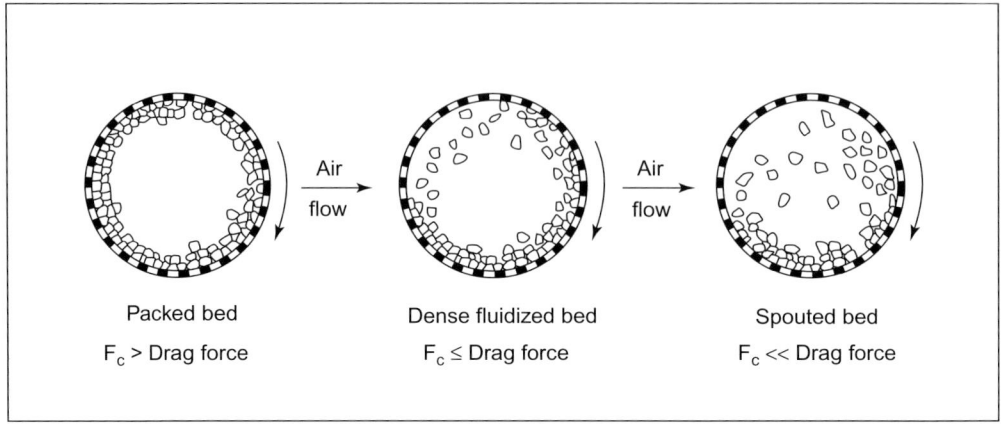

Packed bed   Dense fluidized bed   Spouted bed

F_c > Drag force $F_c \leq$ Drag force $F_c \ll$ Drag force

Figure 1.21   Fluidization regimes in a centrifugal fluidized bed. From Hanni *et al.* (1976), by permission of the Institute of Food Technologists, USA.

diameter 0.25 m and length 2.5 m, was perforated with 2.4 mm diameter holes giving an open area of 45% and rotated at speeds up to 350 rpm (5.8 Hz). The plenum was divided into two equal lengths, allowing different drying conditions in the feed zone and the discharge zone of the bed respectively. Air passed through the perforations of the drying cylinder and left the plenum to be reheated and recirculated. Mechanical details of sealing the air flow and the control of air velocity are also given.

Hanni *et al.* (1976) used the centrifugal fluidized bed for drying diced vegetables of approximately 10 mm in size with the wet feed being blown into the fluidized bed with air. The perforated surface of the cylinder acted as the distributor plate and the centrifugal action kept the particle bed in place close to the cylinder wall. They suggested that the feed end of the bed was characterised by dense-phase fluidization with partially dried material being displaced by fresh feed and passing on to the second half or discharge end of the cylinder. Here the particle bed approached spouted bed conditions before the particles were discharged radially from the bed by pneumatic transport (Figure 1.21).

Particulate fluidization

Particulate fluidization, where the fluidizing medium is usually a liquid, is characterised by a smooth expansion of the bed. Liquid-solid fluidized beds are used in continuous crystallisers, as bioreactors in which immobilised enzyme beads are fluidized by the reactant solution and in physical operations such as the washing and preparation of vegetables. The empirical Richardson–Zaki equation (Richardson

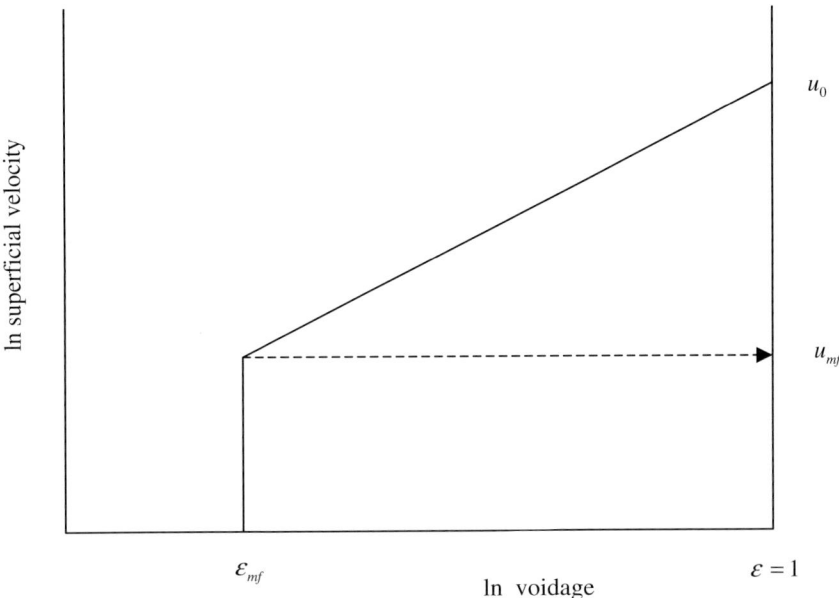

Figure 1.22 Richardson–Zaki plot: measurement of minimum fluidizing velocity for particulate fluidization.

and Zaki, 1954) describes the relationship between the superficial fluid velocity and bed voidage

$$\frac{u}{u_0} = \varepsilon^n \qquad 1.67$$

where u_0 is the superficial velocity at a voidage of unity and n is an index which depends on the particle Reynolds number. Figure 1.22 shows the form of a logarithmic plot of velocity against voidage. The vertical line represents the packed bed where the bed voidage does not change with velocity. The inclined section of the plot represents the fluidized bed, described by equation 1.67, and has a gradient equal to n. The minimum fluidizing velocity is found at the intersection of the two lines. The value of n is usually in the range 2.4–5.0, with smaller values at higher Reynolds numbers. There is a degree of similarity between the fluidization of a particle and the settling of a particle in a fluid. For sedimentation $u_0 = u_t$, i.e. the velocity at a voidage of unity is equal to the terminal falling velocity. However, in particulate fluidization a wall effect exists and

$$\ln u_0 = \ln u_t - \frac{d}{D_{bed}} \qquad 1.68$$

Table 1.2   Richardson–Zaki index as a function of Reynolds number.

$Re < 0.2$	$n = 4.65 + 20\dfrac{d}{D_{bed}}$
$0.2 < Re < 1$	$n = \left(4.4 + 18\dfrac{d}{D_{bed}}\right)Re^{-0.03}$
$1 < Re < 200$	$n = \left(4.4 + 18\dfrac{d}{D_{bed}}\right)Re^{-0.1}$
$200 < Re < 500$	$n = 4.4\,Re^{-0.1}$
$Re > 500$	$n = 2.4$

where D_{bed} is the bed diameter. Thus as the bed diameter increases relative to the particle diameter, fluidization behaviour approaches that of a sedimenting particle. Richardson and Zaki (1954) proposed a series of empirical equations for the relationship between n and Reynolds number; these are listed in Table 1.2. For non-spherical particles, for example vegetable pieces, the value of n will be higher than for spheres; here Richardson and Zaki suggested

$$n = 2.7K''^{0.16} \quad \text{(for } Re > 500\text{)} \tag{1.69}$$

where the shape factor K'' is defined by equation 1.15.

The Richardson–Zaki equation has been found to agree with experimental data over a wide range of conditions. Equally, it is possible to use a pressure drop–velocity relationship such as Ergun to determine minimum fluidization velocity, just as for gas-solid fluidization. An alternative expression, which has the merit of simplicity, is that of Riba *et al.* (1978)

$$Re_{mf} = 1.54 \times 10^{-2} Ga^{0.66}\left(\frac{\rho_s - \rho_f}{\rho_f}\right)^{0.7} \tag{1.70}$$

which is valid in the Reynolds number range $10 < Re < 1000$.

Nomenclature

A	bed cross-sectional area
c_D	drag coefficient
d	mean particle diameter
d_B	bubble diameter
d_e	equivalent spherical diameter
d_s	diameter of a sphere with the same surface area as particle
d_p	diameter of a circle having an area equal to the projected area of particle in its most stable position

d' equivalent diameter of void spaces

D_{bed} bed diameter

D_i diameter of spouted bed gas inlet

F drag force

Fr Froude number

g acceleration due to gravity

Ga Galileo number

H bed height

H_{mf} bed height at minimum fluidization

H_S minimum bed height for slugging

H_{sm} maximum spoutable bed depth

k constant in equation 1.6

K Kozeny's constant

K' volume shape factor

K'' shape factor defined by equation 1.15

L' equivalent length of void spaces

m particle mass

m' mass of displaced fluid

n index in Richardson–Zaki equation

N number of particles of size x

p parameter representing characteristic of a particle size distribution

q parameter representing characteristic of a particle size distribution

Q total volumetric gas flow rate

Q_B volumetric bubble flow rate

Q_{mf} volumetric gas flow rate at minimum fluidization

Re Reynolds number

Re_t Reynolds number at terminal falling velocity

S specific surface

S_B particle surface area per unit bed volume in contact with fluid

t time; average particle circulation time

u superficial velocity; velocity through aperture of distributor plate; relative velocity between sphere and fluid

u_B bubble rise velocity

u_{mb} minimum bubbling velocity

u_{mf} minimum fluidizing velocity

u_{ms} minimum spouting velocity

u_{SB} rise velocity of a single bubble

u_t terminal falling velocity

u' interstitial velocity

u_0 superficial velocity at a voidage of unity

V particle volume

x particle size

x_l lower limit of size distribution
x_u upper limit of size distribution
$x_{3,2}$ surface–volume mean diameter

Greek symbols

α coefficient in generalised Ergun equation
β coefficient in generalised Ergun equation
ΔP bed pressure drop
ΔP_d pressure drop across distributor plate
ε interparticle voidage
μ viscosity
ρ_B bulk density
ρ_f gas density
ρ_s particle density
ϕ sphericity
ω mass fraction of particles of a given size

Subscripts

f fluid
mf minimum fluidizing conditions
s solid

References

Abrahamson, A.R. and Geldart, D., Behaviour of gas-fluidized beds of fine powders part I. Homogeneous expansion, *Powder Tech.*, **26** (1980) 35–46.

Allen, T., Particle size measurement, Chapman and Hall, London, 1981.

Arjona, J.L., Rios, G.M. and Gibert, H., Two new techniques for quick roasting of coffee, *Lebensmittel Wissenschaft Technologie*, **13** (1980) 285–290.

Baeyens, J. and Geldart, D., An investigation into slugging fluidized beds, *Chem. Eng. Sci.*, **29** (1974) 255–265.

Botterill, J.S.M., Fluid-bed heat transfer, Academic Press, London, 1975.

Carlson, R.A., Roberts, R.L. and Farkas, D.F., Preparation of quick-cooking rice products using a centrifugal fluidized bed, *J. Food Sci.*, **41** (1976) 1177–1179.

Couderc, J.-P., Incipient fluidization and particulate systems, in: Davidson, J.F., Clift, R. and Harrison, D., (eds.), Fluidization, 2nd ed., Academic Press, London, 1985.

Davidson, J.F. and Harrison, D., Fluidised particles, Cambridge University Press, Cambridge, 1963.

Davidson, J.F. and Harrison, D., (eds.), Fluidization, Academic Press, London, 1971.

Davidson, J.F., Clift, R. and Harrison, D., (eds.), Fluidization, 2nd ed., Academic Press, London, 1985.

Epstein, N., Applications of liquid-solid fluidization, *Int. J. Chem. Reactor Eng.*, **1** (2003) 1–16.

Geldart, D., Types of gas fluidization, *Powder Tech.*, **7** (1973) 285–292.

Geldart, D., Mixing in fluidized beds, in: Harnby, N., Edwards, M.F. and Nienow, A.W., (eds.), Mixing in the process industries, Butterworth-Heinemann, Oxford, 1992.

Gibilaro, L.G., Fluidization-dynamics: the formulation and applications of a predictive theory for the fluidized state, Butterworth-Heinemann, Oxford, 2001.

Hanni, P.F., Farkas, D.F. and Brown, G.E., Design and operating parameters for a centrifugal fluidized bed drier, *J. Food Sci.*, **41** (1976) 1172–1176.

Hiby, J.W., Untersuchungen über den kritischen Mindestdruckverlust des Anströmbodens bei Fluidalbetten (Fließbetten) [Examination of the critical minimum pressure loss in a fluidised bed], *Chem. Ing. Techn.*, **36** (1964) 228–229.

Jackson, A.T. and Lamb, J., Calculations in food and chemical engineering, Macmillan, Basingstoke, 1981.

Jonke, A.A., Petkus, E.J., Loeding, J.W. and Lawroski, S., Calcination of dissolved nitrate salts, *Nucl. Sci. Eng.*, **2** (1957) 303–319.

Jowitt, R., Heat transfer in some food processing applications of fluidisation, *Chem. Engnr.*, **November** (1977) 779–782.

Kunii, D. and Levenspiel, O., Fluidization engineering, Butterworth-Heinemann, Oxford, 1991.

Leva, M., Fluidization, McGraw-Hill, New York, 1959.

Mathur, K.B., Spouted beds, in: Davidson, J.F. and Harrison, D., (eds.), Fluidization, Academic Press, London, 1971.

Mathur K.B. and Epstein, N., Spouted beds, Academic Press, New York, 1974.

Mathur, K.B. and Gishler, P.E., A technique for contacting gases with coarse solid particles, *A.I.Chem.E.J.*, **1** (1955) 157–164.

Mishra, I.M., El-Temtamy, S.A. and Schugerl, K., Growth of *Saccharomyces cerevisiae* in gaseous fluidized beds, *Eur. J. Appl. Microbiol. Biotechnol.*, **16** (1982) 197–203.

Mugele, R.A. and Evans, H.D., Droplet size distribution in sprays, *Ind. Eng. Chem.*, **43** (1951) 1317–1324.

Mullin, J.W., Crystallization, 3rd ed., Butterworth-Heinemann, Oxford, 1993.

Rankell, A.S., Scott, M.W., Lieberman, H.A., Chow, F.S. and Battista, J.V., Continuous production of tablet granulations in a fluidized bed II. Operation and performance of equipment, *J. Pharm. Sci.*, **53** (1964) 320–324.

Riba, J.P., Routie, R. and Couderc, J.P., Conditions minimales de mise en fluidisation per une liquide [Minimum conditions for initiation of fluidisation with a liquid], *Can. J. Chem. Eng.*, **56** (1978) 26–34.

Richardson, J.F., Incipient fluidization and particulate systems, in: Davidson, J.F. and Harrison, D., (eds.), Fluidization, Academic Press, London, 1971.

Richardson, J.F. and Zaki, W.N., Sedimentation and fluidisation. Part 1, *Trans. Inst. Chem. Engrs.*, **32** (1954) 35–52.

Rios, G.M., Gibert, H. and Baxerres, J.L., Factors influencing the extent of enzyme inactivation during fluidized bed blanching of peas, *Lebensmittel Wissenschaft Technologie*, **11** (1978) 176–180.

Rios, G.M., Gibert, H. and Baxerres, J.L., Potential applications of fluidisation to food preservation, in: Thorne, S., (ed.), Developments in food preservation, volume 3, Elsevier, Amsterdam, 1985, 273–304.

Rowe, P.N., Experimental properties of bubbles, in: Davidson, J.F. and Harrison, D., (eds.), Fluidization, Academic Press, London, 1971.

Rowe, P.N., Estimation of solids circulation rate in a bubbling fluidised bed, *Chem. Eng. Sci.*, **28** (1977) 979–980.

Saxena, S.C. and Vogel, G. J., The measurement of incipient fluidisation velocities in a bed of coarse dolomite at temperature and pressure, *Trans. Inst. Chem. Engrs.*, **55** (1977) 184–189.

Schiller, L. and Naumann, A.Z., Uber die grundlegenden Berechnungen bei der Schwerkraftaufbereitung [A fundamental drag coefficient correlation], *Z. Ver. Deut. Ing.*, **77** (1933) 318–320.

Sherrington, P.J. and Oliver, R., Granulation, Heyden, London, 1981.

Shi, Y.F. and Fan, L.T., Effect of distributor to bed resistance ratio on uniformity of fluidization, *A.I.Ch.E.J.*, **30** (1984) 860–865.

Shilton, N.C. and Niranjan, K., Fluidization and its applications to food processing, *Food Structure*, **12** (1993) 199–215.

Siegel, R., Effect of distributor plate-to-bed resistance ratio on onset of fluidized bed channelling, *A.I.Ch.E.J.*, **22** (1976) 590–592.

Smith, P.G., Introduction to food process engineering, Kluwer Academic, New York, 2003.

Smith, P.G. and Nienow, A.W., Particle growth mechanisms in fluidised bed granulation – part I. The effect of process variables, *Chem. Eng. Sci.*, **38** (1983) 1223–1231.

Stokes, G.G., On the effect of the internal friction of fluids on the motion of pendulums, *Trans. Cambridge Phil. Soc.*, **9** (1851) 8–106.

Vinter, H., Aqueous (and organic) film-coating by fluidisation technology, Proceedings of an International Conference on Fluidisation Technology for Pharmaceutical Manufacturers, Powder Advisory Centre, London, 1982, 1–11.

Wen, C.J. and Yu, Y.H., A generalized method for predicting the minimum fluidization velocity, *A.I.Ch.E.J.*, **12** (1966) 610–612.

Wilhelm, R.H. and Kwauk, M., Fluidization of solid particles, *Chem. Eng. Prog.*, **44** (1948) 201–218.

Yates, J.G., Fundamentals of fluidized-bed chemical processes, Butterworth, London, 1983.

Yerushalmi, J. and Avidan, A., High-velocity fluidization, in: Davidson, J.F., Clift, R. and Harrison, D., (eds.), Fluidization, 2nd ed., Academic Press, London, 1985.

Zenz, F.A., Regimes of fluidized behaviour, in: Davidson, J.F. and Harrison, D., (eds.), Fluidization, Academic Press, London, 1971.

Chapter 2
Characteristics of Aggregative Fluidization

Heat transfer

Two of the most important characteristics of gas-solid fluidized beds are the high rates of heat and mass transfer that are possible and the ability to provide rapid particle mixing. It is these characteristics which are responsible for the use of fluidized beds in drying, freezing, granulation and other operations. In most food-related applications heat is added to the bed by fluidizing the particles with hot inlet gas, although where a fluidized bed is used as a freezer, the inlet air is at a temperature well below the freezing temperature of the food. However, it is also possible to add heat via heating elements in the bed wall or from heat exchanger coils immersed in the bed itself. Similarly, heat can be removed from the bed by using immersed coils in which a coolant circulates.

In considering heat transfer in gas-solid fluidization it is important to distinguish between, on the one hand, heat transfer between the bed and a heat transfer surface (be it heated bed walls or heat transfer coils in the bed) and, on the other hand, heat transfer between particles and the fluidizing gas. Much of the fluidization literature is concerned with the former because of its relevance to the use of fluidized beds as heterogeneous chemical reactors. Gas-particle heat transfer is rather more relevant to the food processing applications of fluidization such as drying, where the transfer of heat from the inlet gas to the wet food particle is crucial.

Correlations for heat transfer coefficients

Correlations for heat transfer coefficients generally take the form

$$Nu = f(Re, Pr) \qquad 2.1$$

The Nusselt number Nu contains the heat transfer coefficient h and is defined as

$$Nu = \frac{hL}{k} \qquad\qquad 2.2$$

where L is a linear dimension which is characteristic of the heat transfer geometry, for example the diameter of a sphere, and k is the thermal conductivity. The Nusselt number is a kind of dimensionless heat transfer coefficient and represents the ratio of the actual rate of heat transfer to that due to conduction alone. The Prandtl number defined by

$$Pr = \frac{c_p \mu}{k} \qquad\qquad 2.3$$

is solely a function of the physical properties of the fluid and is equal to the ratio of kinematic viscosity (or kinematic diffusivity) $\dfrac{\mu}{\rho}$ to thermal diffusivity $\dfrac{k}{\rho c_p}$. It may be thought of as a measure of the relative ability of a fluid to transfer momentum and heat. Based on experimental observation, equation 2.1 takes the form

$$Nu = CRe^a Pr^b \qquad\qquad 2.4$$

where C is a constant and a and b are indices all of which depend upon geometry and the nature of the heat transfer application.

Bed-surface heat transfer

The heat transfer coefficient h_w which characterises heat transfer between the bed wall or an immersed surface and the fluidized bed is defined by

$$q = h_w \Delta T \qquad\qquad 2.5$$

where q is the heat flux and ΔT is the temperature difference between the wall, or surface, and the bed.

The high rates of heat transfer obtainable are due to a number of reasons. First, the presence of particles in a fluidized bed increases the heat transfer coefficient by up to two orders of magnitude, compared with the value obtained with gas alone at the same velocity. This is because the particles tend to reduce the thickness of the boundary layer at the heat transfer surface (Jowitt, 1977). The bed particles are responsible for the transfer of heat and, because of the high rate of particle movement (and very short residence times close to the heat transfer surface), the bulk bed temperature is uniform and comes very close to that of the heat transfer surface. Despite this, Botterill (1975) suggests that values of h_w are limited to about $400\,W\,m^{-2}\,K^{-1}$ and that in many industrial applications fall as low as $60\,W\,m^{-2}\,K^{-1}$. Kunii and Levenspiel (1991) quote experimental data for both vertical and horizontal surfaces in the range 200–$400\,W\,m^{-2}\,K^{-1}$ with higher coefficients at higher fluidizing gas velocities. The heat transfer coefficient appears to decrease with increasing particle size down to approximately $100\,W\,m^{-2}\,K^{-1}$ at a diameter of 2 mm. Kunii and Levenspiel (1991) proposed the correlation

$$\frac{Nu}{1-\delta} = 5 + 0.05\,RePr \qquad\qquad 2.6$$

where δ is the volumetric bubble fraction in the bed, for particles up to 4 mm in diameter; substitution of realistic values into equation 2.6 results in similar values of h_w to those already quoted.

The second reason for high rates of heat transfer is that the volumetric particle heat capacity is about 1000 times greater than that of a gas and therefore approximates to that of a liquid. As Botterill (1975) has pointed out, a fluidized bed is effectively a fluid of high heat capacity but very low vapour pressure. Third, the very high specific surface of the particles results in high heat fluxes.

Gas-particle heat transfer

In applications where heat is added to a fluidized bed by fluidizing the particles with hot gas, the temperature drop between the inlet gas and the bed takes place over only a few particle diameters immediately above the gas distributor plate. Temperatures throughout the bed are uniform, because of the rapid particle mixing, and thus close bed temperature control is possible. Equally, it is possible to use high inlet temperatures without exposing temperature-sensitive food particles to thermal damage. However, significant temperature gradients have been observed where a large heat sink exists in the bed, for example due to the evaporation of liquid during fluidized bed granulation (Smith and Nienow, 1982). Gas-particle heat transfer coefficients are not very large and according to Botterill (1975) can be as low as $20\,W\,m^{-2}\,K^{-1}$. However, this is compensated for by the very high surface area of the bed particles. For spherical particles, the specific surface is equal to $S = \dfrac{6}{d}$ and therefore for a particle diameter of $200\,\mu m$ $S = 3 \times 10^4\,m^2\,m^{-3}$ and for $d = 2\,mm$ $S = 3 \times 10^3\,m^2\,m^{-3}$, although this will be somewhat lower for very large food particles such as peas or beans.

Gas-particle heat transfer coefficient

For a single sphere in a stagnant environment, i.e. where there is no convection, the limiting value of the Nusselt number can be shown (see, for example, Kay and Nedderman, 1985) to be

$$Nu = 2 \qquad\qquad 2.7$$

This represents the lowest possible value of the film heat transfer coefficient when no natural or forced convection currents are present. When there is a relative velocity between the sphere and the surrounding fluid the heat transfer coefficient will increase and heat transfer will improve. Thus for heat transfer to and from spherical particles, correlations take the form of equation 2.8

$$Nu = 2 + CRe^a Pr^b \qquad\qquad 2.8$$

where the second term represents convective heat transfer. In cases where there are significant convection currents the term $Nu = 2$ becomes negligible in comparison and is often omitted.

Kunii and Levenspiel (1991) summarised the experimentally determined values of the gas-particle heat transfer coefficient h_p and the data are shown in Figure 2.1. At $Re > 100$ the value of Nu lies between that for a single particle and a fixed bed and these authors suggest that the relevant correlations are those due to Ranz, equations 2.9 and 2.10 respectively:

Single particle: $Nu = 2 + 0.6Re^{0.50}Pr^{0.33}$ 2.9

Fixed bed: $Nu = 2 + 1.8Re^{0.50}Pr^{0.33}$ 2.10

However, for $Re < 10$ the experimental values of Nu fall sharply with decreasing Reynolds number, well below the theoretical minimum of $Nu = 2$. This is attributable in part to experimental difficulties, for example the problem of measuring particle temperature, and in part to the theoretical interpretation of the data. Botterill (1975) posed the question of what exactly is measured by a bare wire thermocouple inserted in a fluidized bed. Despite the uncertainties in the experimental evidence, Botterill concluded that it probably does indeed measure the particle temperature. This was the assumption of Smith and Nienow (1982) who used bare wire thermocouples to measure bed particle temperatures during fluidized bed granulation. In the region $Re < 10$, as Kunii and Levenspiel (1991) indicate, the data can be represented by an expression due to Kothari

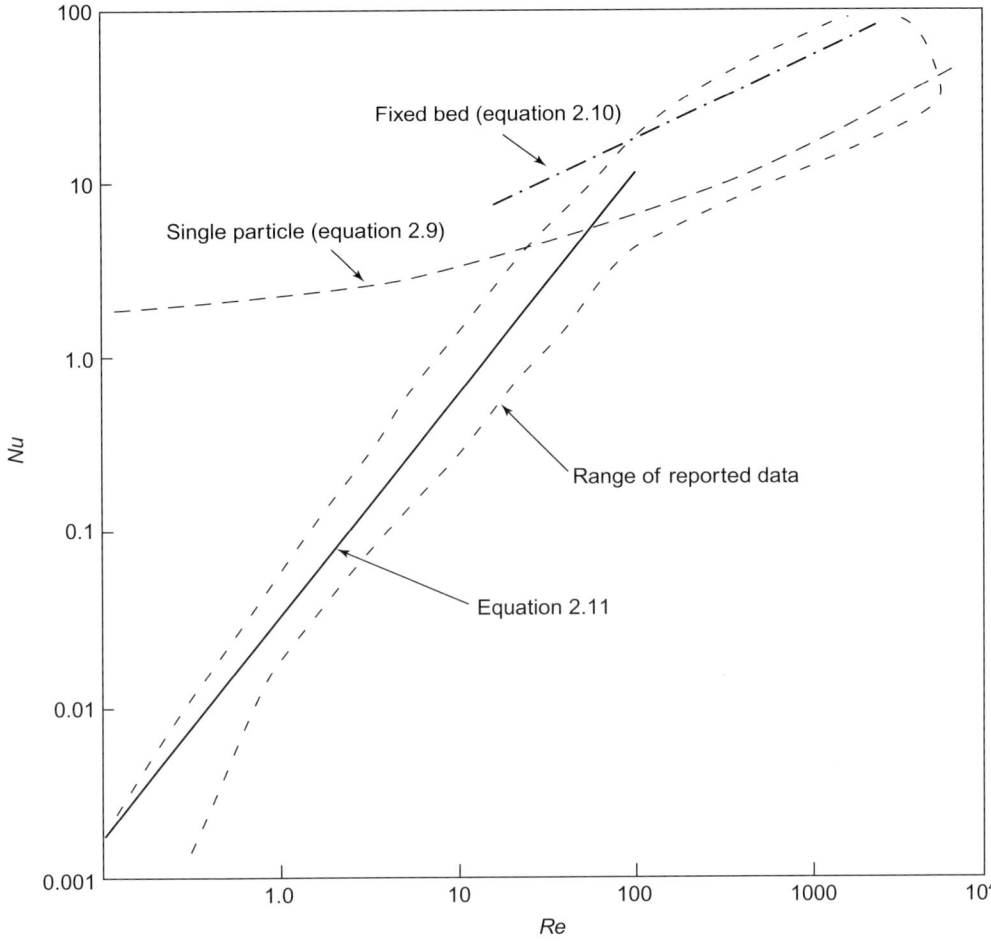

Figure 2.1 Experimental gas-particle heat transfer coefficients. Adapted from Kunii, D. and Levenspiel, O., Fluidization engineering, 1991, with permission from Elsevier.

$$Nu = 0.03Re^{1.3} \qquad\qquad 2.11$$

However, Botterill (1975) suggests that this high dependence on Reynolds number is not consistent with other studies and that, based on the work of Rabinovich with 25 mm particles, a more reasonable index on *Re* would be 0.77.

Kunii and Levenspiel (1991) identify two kinds of heat transfer coefficient to describe gas-particle heat transfer. The coefficient for a single particle, or local coefficient, h_p is that pertaining to a single particle at high temperature T_p introduced suddenly into a bed of cooler particles at a temperature T_{bed} and is defined by

$$\frac{dq}{dt} = mc_p\frac{dT_p}{dt} = h_p(T_p - T_{bed})\qquad\qquad 2.12$$

where q is the heat flux and m the mass of a single particle. Second, an overall or whole-bed coefficient h_{bed} can be defined for the case where a bed of cooler particles is heated by hot inlet gas. As both Kunii and Levenspiel and Botterill explain, in order to interpret the experimental data and calculate the heat transfer coefficient, assumptions must be made about the nature of gas flow through the bed. In fluidized beds of large particles, where the gas cloud around a bubble is minimal, gas flow through the bed is better described by a plug flow model than a well-mixed model. Consequently h_{bed} is approximately equal to h_p and thus the experimental heat transfer coefficients obtained with larger particles (and therefore at larger Reynolds number) are closer to theoretical expectations. However, in beds of fine particles a large proportion of the gas is associated with the bubble cloud and the majority of gas passes through the bed in the bubble phase. Because of the impossibility of tracking individual particles, the experimental coefficients indicated in Figure 2.1 are based on the whole-bed coefficient h_{bed} and therefore on the plug flow assumption. Consequently, the degree of gas-particle contact that was assumed is greater than the actual case for fine particles, with the result that the reported heat transfer coefficients are lower than the actual single particle heat transfer coefficient, that is $h_{bed} < h_p$. As was pointed out earlier, gas-particle heat transfer occurs in a very shallow zone just above the distributor plate where the bubbles are small. However, poor gas distribution at the distributor or 'bubble by-passing' may give lower heat transfer coefficients than are predicted (Xavier and Davidson, 1985) for exactly the same reasons.

An indication of the magnitude of the gas-particle film heat transfer coefficient for both single particle and fixed bed models can be obtained by using equations 2.9 and 2.10 respectively together with data from Smith (2003) for two particles: (i) surface-volume mean diameter = $600\,\mu m$, particle density = $2400\,kg\,m^{-3}$ and (ii) surface-volume mean diameter = $9\,mm$, particle density = $1200\,kg\,m^{-3}$. In each case the Reynolds number at minimum fluidization was calculated from the Ergun equation by assuming a bed voidage at minimum fluidization of 0.45 and the density and viscosity of air to be $1.1\,kg\,m^{-3}$ and $2 \times 10^{-5}\,Pa\,s$ respectively. Table 2.1 shows the results of this example calculation and gives both the Nusselt number and the corresponding heat transfer coefficient assuming, for the gas, $Pr = 1$. At any given Reynolds number the actual value of the heat transfer coefficient might be expected to lie between the two calculated values.

Table 2.1 Example calculation of gas-particle heat transfer coefficients.

d		Single particle (equation 2.9)		Fixed bed (equation 2.10)	
		Nu	h (W m^{-2}K^{-1})	Nu	h (W m^{-2}K^{-1})
$600\,\mu$m	$Re_{mf} = 12$	4.1	170	8.2	343
	$Re = 36$	5.6	233	12.8	533
	(i.e. $Re = 3Re_{mf}$)				
9 mm	$Re_{mf} = 1085$	21.8	60	61.3	170
	$Re = 3255$	36.2	101	104.7	291
	(i.e. $Re = 3Re_{mf}$)				

Mass transfer

Correlations for mass transfer coefficients

For mass transfer in the gas phase, the molar flux of a particular component N (in kmol m^{-2}s^{-1}) is related to the concentration difference in the gas phase ΔC, expressed in terms of molar concentration (kmol m^{-3}), by

$$N = k_g \Delta C \qquad\qquad 2.13$$

where the film mass transfer coefficient k_g has units of m s^{-1}. Film mass transfer coefficients can be correlated with physical properties and process variables in a manner which is analogous to that for heat transfer. The Sherwood number defined by

$$Sh = \frac{k_g L}{D} \qquad\qquad 2.14$$

is the mass transfer equivalent of the Nusselt number and is a measure of the ratio of convective mass transfer, represented by the mass transfer coefficient, to molecular diffusion represented by the diffusivity D. The Schmidt number, the equivalent of the Prandtl number, defined by

$$Sc = \frac{\mu}{\rho D} \qquad\qquad 2.15$$

is the ratio of kinematic viscosity to diffusivity and is therefore a measure of the effectiveness of momentum transfer to mass transfer by diffusion. Again, analogous to heat transfer, the mass transfer coefficient which describes the diffusion of a volatile component from a sphere, when placed in a stagnant environment where there is no convective mass transfer, is given by

$$Sh = 2 \qquad\qquad 2.16$$

indicating the limiting value of Sh for mass transfer from a sphere. The diffusional component of mass transfer is always present but will be dominated by convection in turbulent systems. Thus mass transfer correlations for a particle in a turbulent fluid usually take the form

$$Sh = 2 + CRe^a Sc^b \qquad\qquad 2.17$$

where the second term represents convective mass transfer and the term $Sh = 2$ is often negligible in comparison.

Gas-particle mass transfer

The experimental mass transfer coefficients for gas-particle mass transfer in a fluidized bed (Figure 2.2), as summarised by Kunii and

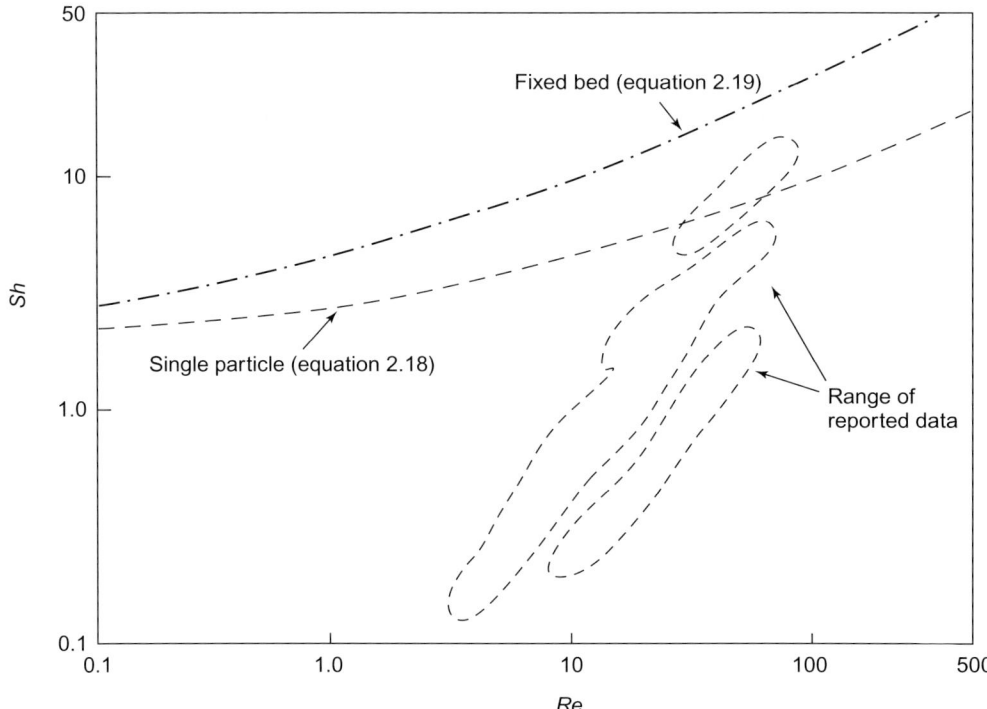

Figure 2.2 Experimental gas-particle mass transfer coefficients. Adapted from Kunii, D. and Levenspiel, O., Fluidization engineering, 1991, with permission from Elsevier.

Levenspiel (1991), show a similar pattern to that for heat transfer, as might be expected because of the analogy between heat and mass transfer. Again, as particle size decreases (and therefore as fluidizing gas velocity and Reynolds number decrease), *Sh* should approach the limiting value of 2 but the experimental data show that *Sh* falls well below this for *Re* < 1. However, at higher Reynolds numbers, above about 100, the data fall between the curves for a single sphere and a fixed bed respectively, the equations for these two cases having the same form as those for heat transfer.

Single sphere: $Sh = 2 + 0.60 Re^{0.50} Sc^{0.33}$ 2.18

Fixed bed: $Sh = 2 + 1.8 Re^{0.50} Sc^{0.33}$ 2.19

This discrepancy can again be explained by proposing two distinct gas-particle mass transfer coefficients. The single particle coefficient is that obtained when mass transfer of a volatile material occurs from a single particle in the bed, with no mass transfer from other particles and with volatile-free gas flowing through the bed. The whole-bed coefficient represents the transfer of mass from all particles in the bed. In beds of large particles, at high *Re*, gas flow through the bed is closer to plug flow and the whole-bed coefficient is approximately equal to the single particle coefficient. In contrast, with fine particles, as with heat transfer, the majority of gas passes through the bed in the bubble phase. Consequently the degree of gas-particle contact is less than that assumed and the reported coefficients are lower than the actual single particle coefficient predicted by *Sh* = 2.

Table 2.2 gives examples of mass transfer coefficients determined from both the single particle and fixed bed models for the evaporation of water from particles of the same diameter and density as in Table 2.1, assuming the diffusivity of water in air to be $3 \times 10^{-5} \, m^2 s^{-1}$. Once

Table 2.2 Example calculation of gas-particle mass transfer coefficients.

d		Single particle (equation 2.18)		Fixed bed (equation 2.19)	
		Sh	k_g (m s^{-1})	*Sh*	k_g (m s^{-1})
600 μm	$Re_{mf} = 12$	3.8	0.188	7.3	0.364
	$Re = 36$	5.1	0.253	11.2	0.558
	(i.e. $Re = 3Re_{mf}$)				
9 mm	$Re_{mf} = 1085$	18.8	0.063	52.3	0.174
	$Re = 3255$	31.0	0.103	89.1	0.297
	(i.e. $Re = 3Re_{mf}$)				

again, the actual film mass transfer coefficient will lie somewhere between the values calculated from equations 2.18 and 2.19 respectively.

Mixing

Introduction

Mixing, either of liquids or of particulate solids, is one of the most common of all operations in the food processing industries, indeed of the process industries in general. Mixing implies the random distribution, throughout a system, of two or more initially separate components. The reasons for mixing are to bring about intimate contact between different species, for example in order for a chemical reaction to occur, or to provide a new property of the mixture which was not present in the original separate components. The former has led to much of the research work on mixing in fluidized beds being concerned with gas mixing, or with gas-solid contacting, as it relates to the use of fluidized beds as chemical reactors; the extensive models which have been proposed in this area are reviewed by van Deemter (1985). An example of providing a new property of the mixture might be the inclusion of a specific proportion in a food mixture of a given component for nutritional purposes or the blending of a solid food product such as flaked cereals, dried vegetables and fruits and even mixed rations for livestock (Lindley, 1991a).

Despite this there are limited applications of fluidized beds being used as mixers *per se* (Geldart, 1992); the food industry tends to choose from a huge variety of devices for the mixing of solids (Anon., 1997) including tumbling mixers, ribbon mixers and vertical screw mixers. Levey and Swientek (1986) describe a so-called fluidized bed mixer used for the blending of powdered food flavours to which liquid components are added. In contrast to a previous process which used a ribbon mixer, and which resulted in segregation of the less dense components at the top of a batch, this device was reported to cut the mixing time from 20 minutes for a 680 kg batch to 6 minutes for a uniformly mixed 1225 kg batch in the new mixer. The mixer is described as a horizontal cylinder with an axial shaft and 'plough shaped mixing elements' rotating at 165 rpm to give the powder a 'pulsating or fluidized bed action'. Liquid ingredients are sprayed into the mixer. No details of fluidizing gas are given and it is doubtful if this represents a fluidized bed as it is generally understood.

It must be emphasised that particle mixing remains an essential consideration in the use of fluidization in other operations such as drying and granulation. The mixing in a fluidized bed of particles of similar physical properties, i.e. of similar size, density and shape, presents little or no difficulty. Perfect mixing has been assumed in many applications such as the fluidized bed drying of tea (Temple and van Boxtel, 1999) and wheat (Giner and Calvelo, 1987) and in the recognition that poor mixing leads to process failure, for example in the use of an aggregative fluidized bed as a solid substrate fermenter (Bauer and Röttenbacher, 1984). However, real industrial mixing problems involve dissimilar materials. Unfortunately, as with many operations involving particles, there is relatively little basic theory to underpin the practical operation of industrially sized mixers, and very little literature which either describes experimental results from the mixing of food solids or gives design information for mixers based on food materials. Much useful information can be drawn from other industrial applications, however.

Mechanisms of solids mixing

A theory for solids mixing is not available as it is for liquids, as is the case generally for the processing of particulate solids. However, a number of mixing mechanisms can be identified and to a limited extent these can be related to mixing equipment. The solids mixing mechanisms which can be identified are diffusion, convection and shear. Particles diffuse under the influence of a concentration gradient in the same way that molecules diffuse. They move by interparticle percolation, i.e. in the void spaces between other particles under the influence of either gravity or, in higher speed mixers, of centrifugal effects. Fick's law can be used to describe this phenomenon.

Convection describes the movement of groups of particles from one place to another within the mixer volume because of the direct action of an impeller or a moving device within the mixer body. As in convection within fluids, this is likely to be a more significant effect than diffusion but diffusional effects will still be present.

The shear mechanism operates when slipping planes are formed within the particulate mass, perhaps because of the action of a blade, which in turn allow particles to exploit new void spaces through which particles can then diffuse.

In addition to these mixing mechanisms, segregation acts against mixing to separate components which have different physical properties. Segregation is usually due to gravitational forces, but is heightened when centrifugal effects are present and occurs when particles have the possibility of falling through the spaces between other particles. The degree of segregation is a function of particle size (with smaller particles being more likely to segregate), density and shape. Thus, larger size differences and larger density differences in a particulate mixture are likely to bring about increased segregation and make mixing more difficult. It is more difficult to quantify the effect of shape, although gross differences in shape are more likely to lead to poor mixing. In a fluidized bed it is important to emphasise that the density difference between particles is more important than size difference in determining the rate and extent of particle segregation.

Mixing in fluidized beds

The small-scale mixing of particles takes place in the bubble wake as a bubble rises up through the bed and this is counteracted by the tendency of particles to fall under gravity. Thus mixing and segregation is an equilibrium process which depends upon conditions in the fluidized bed, the most important of which is the superficial gas velocity (Kunii and Levenspiel, 1991). At very high velocities the mixing effect is dominant and therefore segregation tends not to be present in fast fluidization or in pneumatic conveying. However, in a bed of mixed size particles, at gas velocities close to the minimum fluidizing velocity of the largest particles, such particles may settle out and concentrate in a layer on the distributor plate (Kunii and Levenspiel, 1991). Similarly, in fluidized bed granulation the growth of particles can result in the segregation of large agglomerates at the bottom of the bed (Smith and Nienow, 1983). As Kunii and Levenspiel point out, fluidized beds which are used specifically for mixing are likely to be operated as bubbling beds, close to the minimum fluidizing velocity of at least some of the bed components.

Vertical mixing of solids: the dispersion model

The dispersion model of solids mixing (Kunii and Levenspiel, 1991) is based on Fick's equation which describes unsteady-state mass transfer in one dimension. Thus

$$\frac{\partial C_S}{\partial t} = D_S \frac{\partial^2 C_S}{\partial z^2} \qquad 2.20$$

in which C_S is the concentration of a group of particular particles (e.g. marked or tagged particles in a tracer experiment) at a given height above the distributor z and as a function of time t. This equation assumes that the variations in concentration in the other two orthogonal dimensions x and y are both equal to zero – in other words that

$$\frac{\partial C_S}{\partial x} = 0 \quad \text{and} \quad \frac{\partial C_S}{\partial y} = 0 \qquad 2.21$$

Table 2.3 Typical values of vertical dispersion coefficient D_S as a function of bed diameter.

D_{bed} (m)	D_S (m^{2}s^{-1})
0.1	0.002–0.03
0.3	0.01–0.1
0.6	0.05–0.2
1.5	0.2–0.5
3.0	0.3–1.0

The dispersion model represents mixing in the vertical dimension for relatively tall (large aspect ratio) beds and beds of fine Geldart group A particles and thus the quantity D_S in equation 2.20 is an effective diffusivity of solids in the vertical dimension. Kunii and Levenspiel refer to this as a 'vertical dispersion coefficient' whilst van Deemter (1985) uses the terminology 'vertical solids eddy diffusivity'. Larger values of D_S indicate greater rates of mixing. Summarising the experimental findings of a number of studies, van Deemter (1985) quotes the values of D_S given in Table 2.3 as typical. These data suggest that the effect of bed diameter on vertical mixing is significant. In addition, the dispersion coefficient appeared to increase as the proportion of fine particles in the bed (defined in this instance as particles below $44\,\mu$m in diameter) increased up to about 12% but decreased thereafter; van Deemter (1985) further suggested that the effect of gas velocity on D_S is small. However, in contrast to this observation, Kunii and Levenspiel (1991) proposed equation 2.22, based on experimental work by a number of authors, for small beds of about 0.15 m diameter

$$D_S = 0.06 + 0.1u \qquad 2.22$$

Here the dispersion coefficient (m^{2}s^{-1}) is directly proportional to the superficial gas velocity, over the range 0.1–1 m s^{-1}. For larger diameter beds, in the range 0.1–3 m (and gas velocities between 0.2 and 0.5 m s^{-1}), they suggested

$$D_S = 0.30 D_{bed}^{0.65} \qquad 2.23$$

where D_{bed} is the bed diameter (m). Miyauchi *et al.* (1981) fluidized Geldart group A particles at relatively high gas velocities (u greater than 1 m s^{-1}) and proposed equation 2.24

$$D_S = 0.757 u^{0.5} D_{bed}^{0.9} \qquad 2.24$$

where the units of u and D_{bed} are again m^{2}s^{-1} and m respectively.

In addition to the vertical mixing of solids, a degree of horizontal mixing occurs and this is especially important in long, shallow 'plug flow' beds of the type used in fluidized bed drying or freezing (see Chapter 3).

Rate of mixing

Based upon observations of the rise of particles in the bubble wake and spout (see Chapter 1, p 18), Rowe (1977) showed that the average particle circulation time t around a bed was inversely proportional to the excess gas velocity

$$t = \frac{H_{mf}}{0.6(u - u_{mf})} \qquad 1.8$$

Geldart (1992) suggested a modified form of equation 1.8, thus

$$t = \frac{H_{mf}}{(f_W + 0.38 f_S)Q_B} \qquad 2.25$$

where Q_B is the volumetric bubble flow rate and f_W and f_S are fractional measures of the volumes occupied by particles, compared to the bubble volume, in the bubble wake and spout respectively. Equation 2.25 predicts a slightly longer circulation time than does the Rowe model. Geldart suggests that the batch mixing of non-segregating particles requires between five and ten circulation times or that the mean residence time in continuous mixing should be 5–10 times greater than that given by equation 2.25. According to Schofield (1977), for the same relative gas velocity and for the same mixture in geometrically similar beds, mixing time is proportional to bed diameter to the power 0.6.

Mixing and segregation of dissimilar particles

Mechanisms

Much of the fundamental work on the mixing and segregation of dissimilar particles in gas-solid fluidized beds is due to Rowe, Nienow, Chiba and their co-workers. This work has been summarised by Nienow and Chiba (1985) and, more briefly, by Geldart (1992). Details of the experimental techniques employed, including the model systems employed, the injection of single bubbles into a fluidized bed, the use of two-dimensional beds and the sectioning of the bed and analysis of composition following mixing, are given *inter alia* by Rowe et al. (1965, 1972a, 1972b). This work was concerned with batch mixing, as is indeed much of the mixing in the food industry today, although it was later extended to examine continuous operation (Nienow and Naimer, 1980).

The fluidization of mixtures of particles which differ in size and/or density establishes an equilibrium between mixing and segregation of the components. Segregation is particularly important in the food industry because the materials which are usually mixed together have a very wide range of properties (Lindley, 1991b). For most applications good mixing is required but for others segregation may be desirable, for example the separation of contaminating stones in a batch of seeds (Thomas *et al.*, 1993). Rowe *et al.* (1972a) were the first to propose, for a binary system, a model which identifies one component that tends to sink to the bottom of the bed, the jetsam, and one component that tends to float to the bed surface, the flotsam. Subsequent work by Nienow *et al.* (1978) showed that the same tendencies exist in multi-component mixtures.

The broad pattern of mixing and segregation for a binary system may be summarised as follows. Where the two component particles are of equal density a difference in particle size produces a segregating system, with the larger particles being jetsam. However, differences in particle size of up to an order of magnitude may still result in good mixing in a bubbling bed. There is rather less segregation with a wide particle size distribution of the bed contents than with a clearly bimodal distribution. Geldart (1992) suggests that the segregation of equal density but differently sized particles is significantly worse in fluidized beds which use a porous plate distributor because the bubbles formed at the distributor are too small to lift the larger particles; nozzles and bubble caps are to be preferred. With particles of equal size a density difference in the bed produces segregation more readily, with the denser particles always being the jetsam and forming a layer of pure jetsam at the bottom of the bed; jetsam particles may also be present within the upper part of the bed. For the case of mixtures of small denser particles and large lighter particles, the small denser particles are more likely to be jetsam and will only become flotsam if they are able to percolate through the interstices of the larger particles. Nienow and Chiba (1985) point out that even should this circumstance arise, the small dense particles will revert to being jetsam once the larger particles are fluidized. Further, the denser particles will become jetsam despite having a lower minimum fluidizing velocity.

The classic mechanism of particles being lifted up through the bed in the bubble wake and in the spout behind a bubble (see Chapter 1, p 18) still operates when the bed is composed of two distinct layers: jetsam at the bottom and flotsam above. However, Rowe, Nienow and co-workers also showed that bubbles are responsible for segregation because the denser jetsam particles tend to fall preferentially through the temporarily disturbed region which exists behind a rising bubble.

Patterns of particle segregation

Figures 2.3 and 2.4 indicate the way in which the concentrations of flotsam and jetsam change with increasing gas velocity for each of two cases respectively: a flotsam-rich mixture (with less than 50% by volume of jetsam) and a jetsam-rich mixture (with less than 50% by volume of flotsam). Each figure represents the variation of the mass fraction of jetsam particles x_J with bed height.

Figure 2.3a (at a gas velocity slightly above the lower of the two minimum fluidizing velocities) shows that a layer of pure jetsam develops at the bottom of the bed whilst above this the remainder of the bed contains a lower, but near uniform, concentration of jetsam x'_J. As the fluidizing gas velocity increases the concentration of jetsam in the bed becomes more uniform and approaches the nominal average concentration in the whole bed $\bar{x}_J$ (Figure 2.3c). The degree of mixing in such cases can be quantified using a mixing index M defined as

$$M = \frac{x'_J}{\bar{x}_J}$$ 2.26

Perfect mixing therefore is represented by $M = 1$. Incidentally this definition of mixing index should not be confused with those mixing

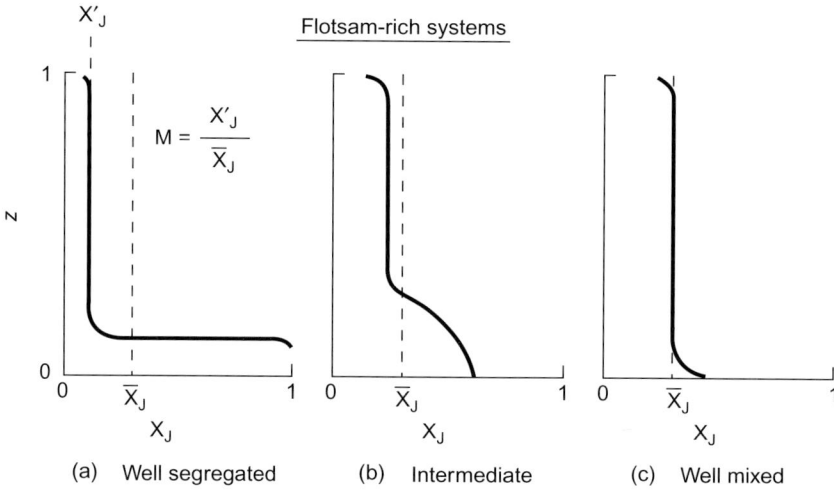

Figure 2.3 Segregation pattern in a flotsam-rich system. Reprinted from Davidson, J.F., Clift, R. and Harrison, D., Fluidization, 2nd ed., Academic Press, 1985, with permission from Elsevier.

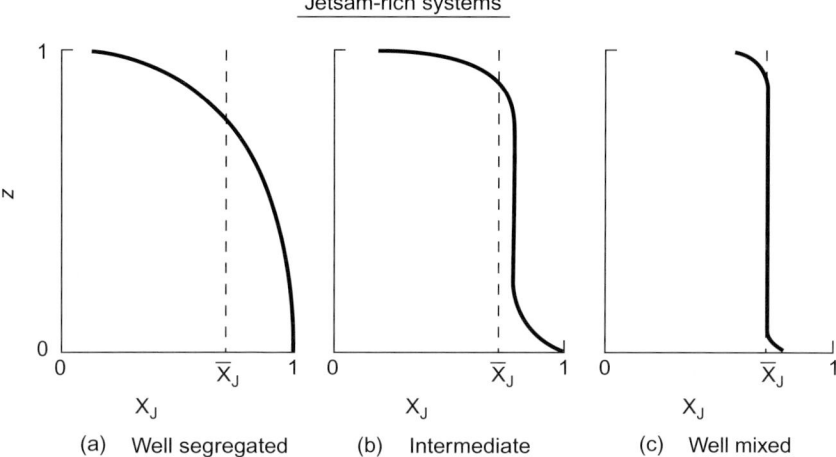

Figure 2.4   Segregation pattern in a jetsam-rich system. Reprinted from Davidson, J.F., Clift, R. and Harrison, D., Fluidization, 2nd ed., Academic Press, 1985, with permission from Elsevier.

indices which are based on standard deviation or variance as the measure of the difference between the measured local concentration and average concentration (Rielly *et al.*, 1994; Smith, 2003) and which are usually used to assess mixer performance.

Nienow *et al.* (1978) showed that the relationship between mixing index and gas velocity could be written as

$$M = \frac{1}{1 + e^{-F}}$$ 2.27

where F is a dimensionless velocity defined by

$$F = \left(\frac{u - u_{TO}}{u - u_F}\right) \exp\left(\frac{u}{u_{TO}}\right)$$ 2.28

In equation 2.28 u_F is the lower of the two minimum fluidizing velocities of the two types of particle in the mixture and u_{TO} is the velocity at which mixing 'takes over' or begins to dominate segregation. Thus, as the superficial gas velocity in the bed is increased, the mixing index increases from $M = 0$ at the lower minimum fluidizing velocity ($u = u_F$), where the bed is quiescent with no particle movement because of the absence of bubbles, to $M = 0.5$ when, by definition, the velocity is equal to u_{TO}. The mixing index approaches a value of unity as the velocity increases still further (Nienow and Chiba, 1985).

The quantity u_{TO} has been correlated with the minimum fluidizing velocity, particle size, sphericity and density of the two kinds of particle; the fraction of jetsam present in the whole bed; and the bed aspect ratio (i.e. the ratio of packed bed height to bed diameter). Thus u_{TO} can be obtained from the purely empirical expression

$$\frac{u_{TO}}{u_F} = \left(\frac{u_P}{u_F}\right)^{1.2} + 0.9\left(\frac{\rho_1}{\rho_2} - 1\right)^{1.1}\left(\frac{\phi_J d_J}{\phi_{FL} d_{FL}}\right)^{0.7}$$
$$- 2.2\sqrt{\bar{x}_J}\left[1 - \exp\left(\frac{-H_0}{D_{bed}}\right)\right]^{1.4} \qquad\qquad 2.29$$

where u_P is the upper of the two minimum fluidizing velocities and the subscripts on density refer to denser (1) and less dense (2) particles respectively. Thus, for a given operating superficial velocity in the bed, the dimensionless velocity F is calculated from equation 2.28 and hence the mixing index from equation 2.27.

Nienow and Chiba (1985) claim good agreement between the model and experimental data obtained with particles in the size range $70\,\mu$m to 5 mm, particularly towards the extremes as M approaches either 0 or 1, these values of mixing index corresponding to the important practical cases of good mixing and clear segregation respectively. The model indicates that higher gas velocities are required to achieve good mixing as the size and density of the jetsam increase. This might be expected. Less expected is that a larger concentration of jetsam in the bed leads to better mixing and that non-spherical particles appear to mix more readily than spheres, although Nienow and Chiba (1985) suggest that this result should be treated with some caution. The effect of aspect ratio can be explained by the increased bubbling frequency, and therefore smaller bubbles, in shallow beds which leads to enhanced segregation.

In jetsam-rich systems the pattern of segregation is quite different from that for flotsam-rich mixtures. As Figure 2.4 shows, there is a much more gradual change in composition with bed height and no layer of pure flotsam to correspond to the layer of jetsam in Figure 2.3a. Once again, increasing the superficial gas velocity improves the degree of mixing. However, there is no equivalent model to predict the mixing index in this case. As Geldart (1992) points out, it is easier to obtain good mixing with a jetsam-rich system; at a gas velocity $0.3\,\mathrm{m\,s^{-1}}$ higher than the minimum fluidizing velocity of the jetsam, these systems are well mixed whereas almost total segregation occurs at the same velocity in a flotsam-rich mixture.

For the case of equal density but differently sized particles fluidized in a bed with a nozzle-type distributor, Geldart (1992) recommends operating the bed at a gas velocity of 1.5 u_{cf}, where u_{cf} is the minimum velocity for complete fluidization, which may be well above the minimum fluidizing velocity found by the usual experimental method.

u_{cf} can be calculated either by weighting the minimum fluidizing velocity for a given particle size in the mixture u_{mfi} with the mass fraction of that size ω, thus

$$u_{cf} = \sum \omega u_{mfi} \qquad\qquad 2.30$$

or by using instead the minimum fluidizing velocity based on the average size of particles in the upper 10% of the size distribution.

Examples of fluidized bed segregation

Zaltzman *et al.* (1983) described a fluidized bed separator used to remove potatoes from a mixture of stones and clods. The mixture, direct from the field, is fed to a fluidized bed of sand with a fluidized bulk density of approximately $1400\,\mathrm{kg\,m^{-3}}$. The bed consists of two compartments. In the first section the stones and clods, with a density of about $1600\,\mathrm{kg\,m^{-3}}$, sink and effectively form the jetsam before being discharged down an incline into a second fluidized bed. Here they are removed by means of a narrow open-sided drum which rotates in the bed and lifts out stones on perforated flights which are attached to the inner peripheral surface of the drum. The potatoes, which have a density of about $1000\,\mathrm{kg\,m^{-3}}$, are removed as flotsam in the first section. The results of field tests showed that the device can separate potatoes from clods and stones efficiently with a potato recovery of about 99.5% and 100% rejection of the contaminants. This was achieved with feed rates up to $5\,\mathrm{t\,h^{-1}}$ and up to 61.6% potatoes in the feed. Zaltzman and Schmilovitch (1986) described further developments of the technique which increased the capacity to $22\,\mathrm{t\,h^{-1}}$.

In a similar application Thomas *et al.* (1993) developed a process for the separation of stones and sand particles from coriander seeds using a bed fluidized with air. Coriander seeds between 2 and $5\,\mathrm{mm}$ in diameter, and with a minimum fluidizing velocity of $0.8\,\mathrm{m\,s^{-1}}$, were separated from denser stones of a similar size range and from sand particles ranging in size from 0.3 to $1\,\mathrm{mm}$. Fluidization for 5 minutes was sufficient to effect complete separation.

Nomenclature

a index
b index
c_p heat capacity at constant pressure
C   coefficient; molar concentration

C_S concentration of tagged particles

d particle diameter

D diffusivity

D_{bed} bed diameter

D_S effective diffusivity of solids in the vertical dimension

f_W fraction of total bubble volume occupied by particles in the bubble wake

f_S ratio of volume of particles in the spout to bubble volume

F dimensionless gas velocity

h film heat transfer coefficient

h_{bed} overall or whole-bed gas-particle heat transfer coefficient

h_p single particle gas-particle heat transfer coefficient

h_w bed-surface heat transfer coefficient

H_{mf} bed height at minimum fluidization

H_0 packed bed height

k thermal conductivity

k_g film gas mass transfer coefficient

L characteristic length

m mass of a single particle

M mixing index

N molar flux of component A

Nu Nusselt number

Pr Prandtl number

q heat flux

Q_B volumetric bubble flow rate

Re Reynolds number

S specific surface

Sc Schmidt number

Sh Sherwood number

t time; average particle circulation time

T_{bed} bed temperature

T_p particle temperature

u superficial gas velocity

u_{cf} minimum velocity for complete fluidization

u_F lower minimum fluidizing velocity in a binary mixture

u_{mf} minimum fluidizing velocity

u_{mfi} minimum fluidizing velocity for a given particle size

u_P upper minimum fluidizing velocity in a binary mixture

u_{TO} velocity at which mixing dominates segregation

x_J concentration of jetsam

x'_J uniform concentration of jetsam in the upper part of the bed

$\bar{x}_J$ average concentration of jetsam in the whole bed

z height above the distributor

Greek symbols

δ volumetric bubble fraction in the bed
ΔC concentration difference
ΔT temperature difference between the bed wall and fluidized bed
μ viscosity
ρ density
ϕ sphericity
ω mass fraction of particles of a given size

Subscripts

1   denser
2   less dense
FL   flotsam
J   jetsam

References

Anon., Blending basics, *Food Review*, **24** (9) (1997) 39–40.

Bauer, W. and Röttenbacher, L., Regelung der Substratzufuhr eines Gas/Feststoff-Wirbelschichtfermenters über die Bestimmung des Fluidisationsverhaltens [Control of material feed into a gas/solids fluidised bed fermenter on the basis of fluidisation behaviour], *Int. Z. Lebensmittel Tech. Verfahrenstechnik*, **35** (1984) 18–23.

Botterill, J.S.M., Fluid-bed heat transfer, Academic Press, London, 1975.

Geldart, D., Mixing in fluidized beds, in: Harnby, N., Edwards, M.F. and Nienow, A.W., (eds.), Mixing in the process industries, Butterworth-Heinemann, Oxford, 1992.

Giner, S.A. and Calvelo, A., Modelling of wheat drying in fluidized beds, *J. Food Sci.*, **52** (1987) 1358–1363.

Jowitt, R., Heat transfer in some food processing applications of fluidisation, *Chem. Engnr.*, **November** (1977) 779–782.

Kay, J.M. and Nedderman, R.M., Fluid mechanics and transfer processes, Cambridge University Press, Cambridge, 1985.

Kunii, D. and Levenspiel, O., Fluidization engineering, Butterworth-Heinemann, Oxford, 1991.

Levey, V. and Swientek, R.J., Fluidized bed mixer cuts blending time 70%, *Food Processing USA*, **47** (1986) 166–167.

Lindley, J.A., Mixing processes for agricultural and food materials: 1. Fundamentals of mixing, *J. Agric. Eng. Res.*, **48** (1991a) 153–170.

Lindley, J.A., Mixing processes for agricultural and food materials: 3. Powders and particulates, *J. Agric. Eng. Res.*, **49** (1991b) 1–19.

Miyauchi, T., Furwsaki, S., Mooroka, S. and Ikdea, Y., Transport phenomena and reaction in fluidized catalyst beds, *Adv. Chem. Eng.*, **11** (1981) 276–448.

Nienow, A.W. and Chiba, T., Fluidization of dissimilar materials, in: Davidson, J.F., Clift, R. and Harrison, D., (eds.), Fluidization, 2nd ed., Academic Press, London, 1985.

Nienow, A.W. and Naimer, N.S., Continuous mixing of two particulate species of different density in a gas fluidised bed, *Trans. Inst. Chem. Engrs.*, **58** (1980) 181–186.

Nienow, A.W., Rowe, P.N. and Cheung, L.Y.-L., A quantitative analysis of the mixing of two segregating powders of different density in a gas-fluidised bed, *Powder Tech.*, **20** (1978) 89–97.

Rielly, C.D., Smith, D.L.O., Lindley, J.A., Niranjan, K. and Phillips, V.R., Mixing processes for agricultural and food materials: 4. Assessment and monitoring of mixing systems, *J. Agric. Eng. Res.* **59** (1994) 1–18.

Rowe, P.N., Estimation of solids circulation rate in a bubbling fluidised bed, *Chem. Eng. Sci.*, **28** (1977) 979–980.

Rowe, P.N., Nienow, A.W. and Agbim, A.J., The mechanisms by which particles segregate in gas fluidised beds: binary systems of near-spherical particles, *Trans. Inst. Chem. Engrs.*, **50** (1972a) 310–323.

Rowe, P.N., Nienow, A.W. and Agbim, A.J., A preliminary quantitative study of particle segregation in gas fluidised beds: binary systems of near-spherical particles, *Trans. Inst. Chem. Engrs.*, **50** (1972b) 324–333.

Rowe, P.N., Partridge, B.A., Cheney, A.G., Henwood, G.A. and Lyall, E., The mechanisms of solids mixing in fluidised beds, *Trans. Inst. Chem. Engrs.*, **43** (1965) 271–286.

Schofield, C., Scale-up of powder and paste mixers, *Chem. Ind.*, **February** (1977) 105–108.

Smith, P.G., Introduction to food process engineering, Kluwer Academic, New York, 2003.

Smith, P.G. and Nienow, A.W., On atomising a liquid into a gas fluidised bed, *Chem. Eng. Sci.*, **37** (1982) 950–954.

Smith, P.G. and Nienow, A.W., Particle growth mechanisms in fluidised bed granulation – part I. The effect of process variables, *Chem. Eng. Sci.*, **38** (1983) 1223–1231.

Temple, S.J. and van Boxtel, A.J.B., Modelling of fluidized bed drying of black tea, *J. Agric. Engng. Res.*, **74** (1999) 203–212.

Thomas, P.P., Gopalakrishnan, N., Sudhilal, N., Poulose, T.P. and Varghese, E., A simple method for the separation of stones from coriander seeds based on the use of fluidization technique, *J. Food Sci. Tech. India*, **30** (1993) 303–305.

van Deemter, J., Mixing, in: Davidson, J.F., Clift, R. and Harrison, D., (eds.), Fluidization, 2nd ed., Academic Press, London, 1985.

Xavier, A.M. and Davidson, J.F., Heat transfer in fluidized beds, in: Davidson, J.F., Clift, R. and Harrison, D., (eds.), Fluidization, 2nd ed., Academic Press, London, 1985.

Zaltzman, A. and Schmilovitch, Z., Evolution of a potato, fluidized bed medium separator, *Trans.A.S.A.E.*, **29** (1986) 1462–1469.

Zaltzman, A., Feller R., Mizrach, A. and Schmilovitch, Z., Separating potatoes from clods and stones in a fluidized bed medium, *Trans.A.S.A.E.*, **26** (1983) 987–995.

Part Two
Applications

Chapter 3
Freezing

Low-temperature preservation of foods

Introduction

The storage of foodstuffs at low temperature reduces the growth rate of micro-organisms and effectively removes liquid water, which then becomes unavailable to support microbial growth, such that both microbiological and biochemical spoilage are decreased. Freezing has long been used for the preservation of high-value foods such as meat and fish and over the last 40 years its use has been extended very significantly to a wide range of convenience foods including prepared vegetables, cakes, pastries and desserts, and complete ready meals. Often, as in the case of peas and other vegetables, freezing has grown in importance because the quality of the frozen product is significantly better than that produced using more traditional thermal preservation methods. Consequently, freezing is used only for high-quality raw materials and for high-value manufactured foods.

In general, the freezing and thawing of foods, if properly controlled, has little or no effect on the nutritional value of frozen foods and allows considerable retention of organoleptic properties such as taste, colour and texture. Where adverse effects do occur, for example the loss of texture, this is often associated with slow freezing rates and the formation of large ice crystals which result in structural damage to cells, the inability of the food to retain water and high drip loss. Poor temperature control and subsequent partial thawing and recrystallisation can lead to the same problem. There is also the potential for high rates of vitamin loss during prolonged storage and oxidative rancidity may occur during the frozen storage of foods with a high fat content.

It is important at this point to make the distinction between the terms 'freezing' and 'frozen storage'. The former may be regarded as part of the primary manufacturing process and uses, for example, fluidized bed freezers or plate freezers depending upon the nature of the food to be frozen. Frozen storage refers to the storage of food at

temperatures usually between −18°C and −30°C, with lower temperatures generally resulting in a longer shelf-life. Domestic 'freezers', in
reality frozen storage cabinets, usually operate at −18°C. Both domestic
freezers and industrial cold stores are designed not to freeze food but
to store food which is already frozen at a suitably low temperature.
Thus, in the context of industrial production, food should leave the
primary freezing plant (and enter the cold store) at its frozen storage
temperature.

In a fluidized bed freezer the food solids to be frozen are fluidized
by refrigerated air at temperatures of −40°C or below. Holdsworth
(1987) suggests that fluidized bed freezing and air blast freezing are
the two most important freezing methods. Jowitt (1977) describes
freezing as the most significant of the food applications of fluidization
in which the product is individually quick frozen (IQF), i.e. it is free
flowing, undamaged and attractively glazed. Products frozen in a fluidized bed are easier to handle subsequent to the freezing stage of the
process (Holdsworth, 1983) and suffer lower weight loss than those
frozen in a continuous moving belt blast freezer (Fennema *et al.*, 1973)
whilst at the same time they are true IQF products (Canet, 1989).
Persson (1967) suggested that mass loss is typically less than 1%.
Fluidized bed freezing is generally reported as being suitable for uniformly shaped food pieces of diameter up to about 40 mm (Canet, 1989)
and has found wide application in the freezing of fruits and vegetables.
The high heat transfer rate obtainable in a fluidized bed is the major
reason for its exploitation in food freezing (Sheen and Whitney, 1990).
Applications of fluidization to food freezing are almost exclusively
concerned with the use of gas-solid fluidized beds, although Fikiin
(1992) reports a technique in which a liquid 'fluidized bed' is used for
the chilling of whole fish.

Industrial freezing equipment

The equipment used for particular freezing applications is determined
in part by the physical form of the food to be frozen (Smith, 2003). The
plate freezer consists of a series of (usually horizontal) parallel, hollow
extruded aluminium plates through which a refrigerant circulates at
low temperature. The food blocks are placed between the plates which
are then moved together with a hydraulic ram and therefore the plate
freezer is limited to flat foods such as fish blocks and rectangular
shallow packs. A variant is the vertical plate freezer used on board
fishing vessels in which whole fish are packed into the spaces between
the plates and the gaps between the fish are filled with water, resulting
in an ice block containing tightly packed fish. Irregularly shaped foods
of varying size are more suited to the blast freezer in which air, which

has been cooled by indirect contact with a refrigerant in a heat exchanger, is blown over the surface of the food. Blast freezers may take the form of a simple cabinet into which batches of food are wheeled on trolleys. Continuous versions employ conveyor belts to transport food through either a tunnel or a spiral track counter-current to a flow of cold air. The speed of the conveyor is adjusted so that the residence time in the freezer is equal to the required freezing time.

Other freezing techniques include the scraped surface heat exchanger in which a consistently smooth-textured product, for example ice-cream, is achieved by scraping the newly formed solid crystals from the refrigerated internal surface and thus controlling crystal growth. In cryogenic freezing the food to be frozen is first passed through nitrogen vapour at about −50°C and then frozen by direct contact with a spray of liquid nitrogen. In contrast, in immersion freezing the food is immersed directly in a liquid refrigerant such as liquid nitrogen or, with suitable packaging, salt brines, glycerol, glycol or calcium chloride solution.

Fluidized bed freezing

In a fluidized bed freezer it is assumed that the particles are frozen independently and very rapidly to give a free-flowing IQF product with an attractive glazed appearance due to the freezing of water on the particle surface. However, Singh and Heldman (2001) suggest that the product being frozen is not necessarily in suspension at all times and it is common to view fluidized bed freezing as a modification of air blast freezing which exploits the higher heat transfer coefficients encountered in fluidization and therefore allows the use of more compact equipment. Persson and Londahl (1993) quantify this and suggest that the higher heat transfer rates in fluidized freezers result in equipment up to one-third the size of a comparable belt freezer; freezing times are far less than in plate or air blast freezers (Tressler *et al.*, 1968).

Jowitt (1977) suggested that applications of fluidization to food processing in general fall into two groups: first, those where the food particles are themselves fluidized and second, those where the (normally packaged) food is immersed in a fluidized bed of inert solids for the purpose of heating or cooling. In the context of freezing, Ditchev and Richardson (1999) describe the latter as the dynamic dispersion medium (DDM) method.

Fluidized bed freezing was first introduced in the early 1960s (Holdsworth, 1983) for the freezing of peas (Persson and Londahl, 1993) and peas are now used as a reference product in the specification of production rates in commercial freezing equipment (De Michelis and

Calvelo, 1994). Its use expanded rapidly to the freezing, immediately after harvesting, of seasonal crops which can then be stored in bulk, thus optimising both the freezing and packaging operations (Jowitt, 1977). Ditchev and Richardson (1999) place fluidized bed freezing in a historical context and trace the development of industrial freezing techniques from the use of free convection in air to forced convection (i.e. blast freezing) to fluidization and finally to the DDM method.

Holdsworth (1983) suggested that the technique is used mainly for fruit and vegetables and by 1983 reported that over 600 fluidized bed freezer units had been installed world-wide. However, use of fluidized bed freezing has also been reported on dedicated prawn-freezer trawlers where on-board fluidized bed freezers have been used for shrimps and other shellfish (Morrison, 1993). The rapid rate of freezing in a fluidized bed leads to little loss of moisture and therefore both a greater product yield and better product quality due to the covering of a thin ice film around individual food pieces (Tressler *et al.*, 1968). In contrast to an air blast freezer, Persson and Londahl (1993) suggest that air distribution in a fluidized bed is independent of the freezer loading, with no danger of channelling if the bed is only partially loaded. According to Tressler *et al.* (1968), the other advantages of fluidized beds are low initial and installation costs; portability and ease of expansion; and ease of control and hygiene. However, Poulsen (1986) states that, for a capacity of $1000\,kg\,h^{-1}$, the capital cost of a fluidized bed freezer is over twice that of a blast freezing tunnel although the plant floor area is considerably smaller at about $8.5\,m^2$ compared to $50\,m^2$ for the blast freezer.

Industrial fluidized bed freezers are generally operated in continuous mode and various techniques have been adopted to move solids through the bed. These include: gravity feed and a weir at the bed exit (Holdsworth, 1983); a slightly inclined plate to promote the forward flow of solids (Tressler *et al.*, 1968); a distributor taking the form of a moving perforated belt rather than a fixed plate (Holdsworth, 1983); and a vibrating conveyor (Jowitt, 1977). The perforated belt may be manufactured from stainless steel or some form of plastic (Arthey, 1993). According to Arthey (1993), the 'fixed bed' (by which is meant a classic fluidized bed with a fixed distributor plate) results in IQF products because food pieces, for example peas, dropping into cold air cause 'turbulence' which aids the separate freezing of each piece. In contrast, he suggests that this turbulence is unacceptable for delicate products such as broccoli and whole-leaf spinach; a moving belt arrangement is gentler and therefore more suitable. However, Conroy and Ellis (1981) claim that an 'air-cushioning' effect prevents the clustering or bruising of berries when pre-cooled fruit is fed into a fluidized bed.

Ditchev and Richardson (1999) rank each of the developments in food freezing in terms of heat transfer intensity, mass loss from the food and energy consumption. 'Heat transfer intensity' is greatest for DDM and gives the highest heat transfer coefficient (see Heat transfer in fluidized bed freezers, below) whilst the energy consumption is less than in conventional fluidized freezing. However, a particular advantage of the DDM technique is the lower mass loss from the food being frozen. Table 3.1 gives the percentage mass loss for each freezing process at typical air velocities. For example, if granulated ice is used as the inert carrier particle, the increase in the partial pressure of water vapour in the air stream reduces the driving force for the mass transfer of water from the food surface and therefore results in a lower mass loss. The fluidized medium in DDM must be compatible with foodstuffs; in other words, it must be inherently hygienic, have suitable fluidization characteristics and thermal properties and be resistant to thermal shock (Ditchev and Richardson, 1999).

Marin *et al.* (1983) proposed using inert particles with a density of about $1500 \, kg \, m^{-3}$ fluidized at the minimum fluidizing velocity, assuming that most food pieces have a density of approximately $1000 \, kg \, m^{-3}$; consequently the food particles would be buoyant at these operating conditions. These authors report experimental work using either glass beads between 315 and $400 \, \mu m$ in diameter, fluidized at velocities 2–3 times greater than their minimum fluidizing velocity of $0.076 \, m \, s^{-1}$, or small ice particles. Rios *et al.* (1984) suggest that gas velocities greater than twice the minimum fluidizing velocity and a product to bed volume ratio less than 0.4 are required to obtain good contact between the food pieces and the inert particles. Ditchev and Richardson (1999) refer to the experimental freezing of meat packages using both granulated polythene 3.77 mm in diameter and 2 mm diameter granulated ice particles but they do not quote any data. It is suggested that the fluidized immersion method described by Rios and co-workers (Marin *et al.*, 1983; Rios *et al.*, 1984) has the advantage of lower gas velocities than conventional fluidization. However, this is balanced by the possible disadvantages which include the elutriation of smaller inert particles and the need for subsequent separation of gas and particles, and

Table 3.1 Mass loss from food on freezing.

	Air velocity ($m \, s^{-1}$)	% mass loss
Free convection	–	3–4.5
Forced convection	2–5	1.5–3.5
Fluidization	2–5	0.5–1.5
DDM	0.5–2	0.1–0.35

the contamination of food pieces with inert particles, especially with sticky or fissured foods.

Fluidized beds may be operated either alone or in conjunction with other types of freezing plant. Persson and Londahl (1993) suggest that the rapid heat transfer in a fluidized bed allows 'crust' freezing to take place, i.e. freezing of the surface of the food, which effectively separates the bed particles, and that this can be followed by a conventional conveyor belt freezer giving a second freezing zone for completion of the process. Canet (1989) proposed a similar process suitable for larger non-uniform pieces, and which therefore do not give classic fluidized bed behaviour, and for delicate foods such as cauliflower and strawberries. The fluidized bed stage is again responsible for surface or crust freezing, and thus minimal weight loss, followed by a blast freezing stage. The reverse of this was proposed by Fennema *et al.* (1973) in which an air blast is employed as the first stage to freeze the food surface followed by a fluid bed second stage. In a further variation, Mermelstein (1998) reported a two-stage freezer manufactured by BOC Gases in which seafood pieces are first immersed in liquid nitrogen to freeze the outer layer rapidly followed by fluidization with nitrogen vapour. Conroy and Ellis (1981) report that the pre-cooling of fruit is commonly used prior to fluidized bed freezing.

Capacity of fluidized bed freezers

There is very little information in the literature to indicate either the potential capacity of fluidized bed freezing equipment or the required power consumption. Despite a reference by Conroy and Ellis (1981) to the use of (an unquantified) 'deep bed' in fluidized bed freezing, and the suggestion by Persson and Londahl (1993) that fluidized beds are capable of improving the freezing of wet products because a deep bed can accept products with more surface water, it is usual for relatively shallow beds to be employed. Tressler *et al.* (1968) and Fennema *et al.* (1973) both quote depths of about 3 cm for easily fluidizable foods such as peas or corn and 7.5–12.5 cm for partially fluidizable foods such as green beans. Vazquez and Calvelo (1983b), in measuring the residence time distribution in fluidized beds of peas, employed bed depths between 3.4 cm and 6.3 cm. Theoretical support for these observations is provided by the work of Reynoso and Calvelo (1985) who proposed an optimum bed depth between 6 cm and 8 cm based on their analytical model of the production rate of peas in a fluidized freezer (Figure 3.1).

Persson (1967) suggested that optimum operation is a balance between capacity and power consumption. Freezing capacity, expressed

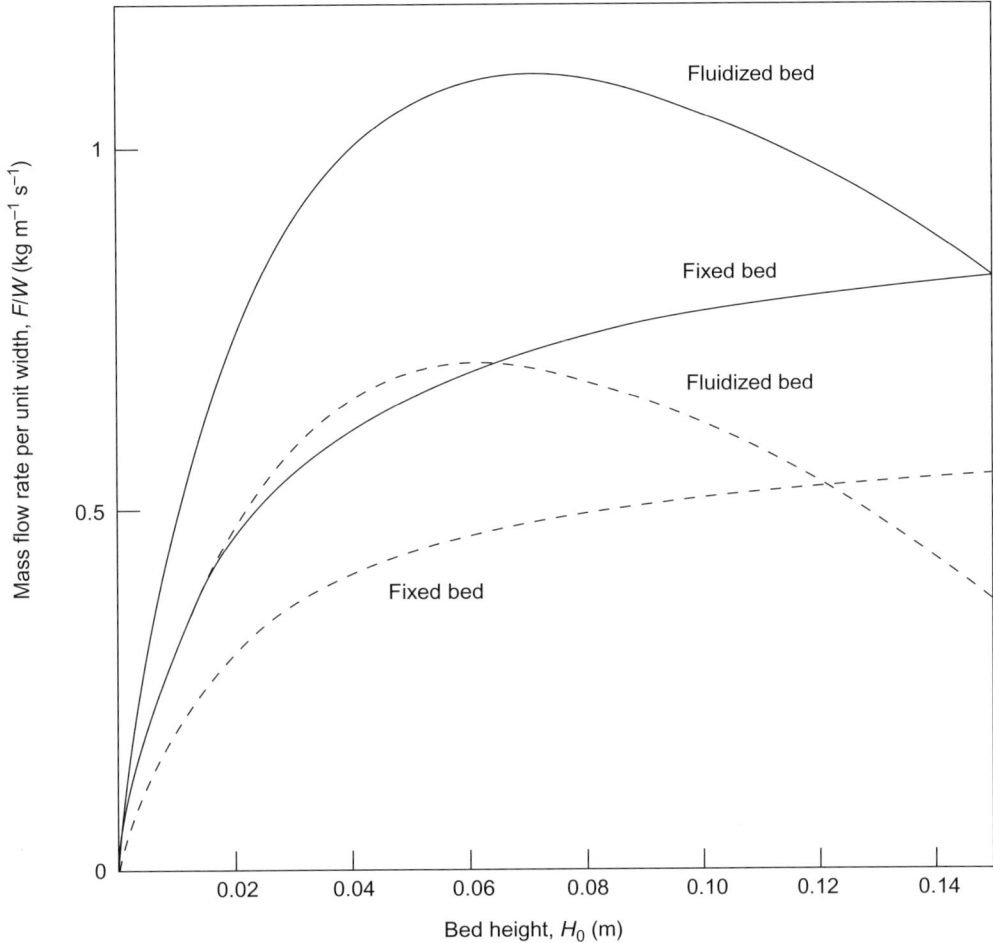

Figure 3.1 Optimum bed depth for fluidized bed freezing: production rate per unit bed width for fixed and fluidized beds at an inlet bed temperature of −30°C (solid line = bed length 6 m, broken line = bed length 4 m). Reprinted from Reynoso, R.O. and Calvelo, A., Comparison between fixed and fluidized bed continuous pea freezers, *Int. J. Refrig.*, **8** (1985) 109–115, with permission from Elsevier.

as mass throughput per unit plate area, increases strongly with bed depth as would be expected. However, throughput per unit of electrical power consumption falls with increasing bed depth (Holdsworth, 1987), although this is not a strong dependence. For the fluidized freezing of peas at an inlet temperature of 15°C and an outlet temperature of −18°C, Persson (1967) presents data (Figure 3.2) which give an optimum bed depth of approximately 13–15 cm, at which the mass throughput per unit plate area is approximately $0.20 \, \text{kg s}^{-1} \text{m}^{-2}$ and the throughput per unit of electrical power consumption is $68 \, \text{kg kWh}^{-1}$ (or $0.019 \, \text{kg kJ}^{-1}$).

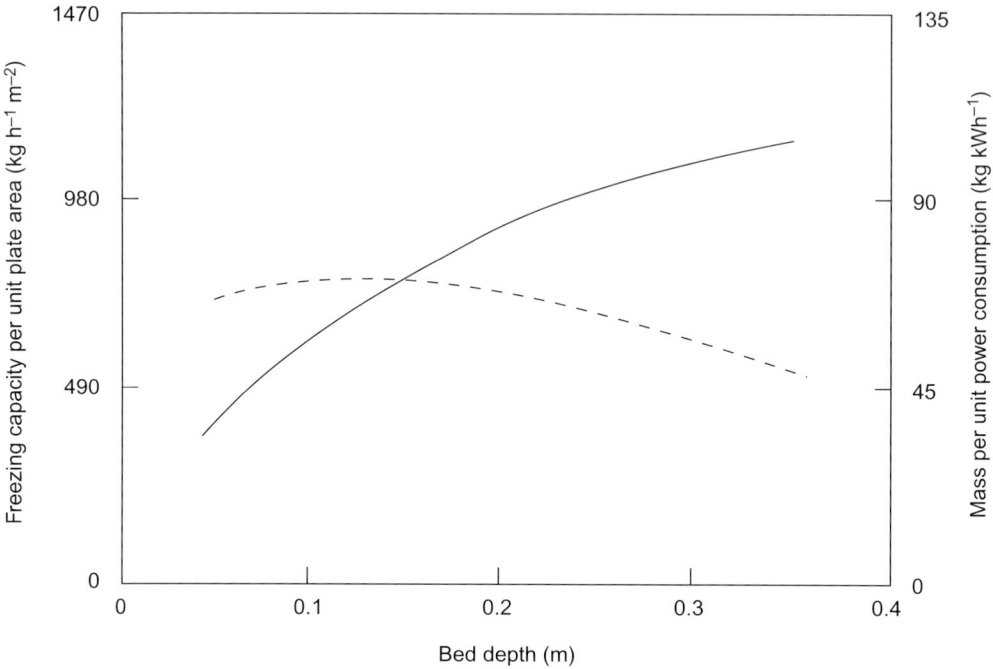

Figure 3.2 Variation of fluidized bed freezer capacity with bed depth (solid line = freezing capacity, broken line = mass per unit power consumption). Adapted from Persson, ASHRAE Journal, June 1967. © American Society of Heating, Refrigerating and Air-Conditioning Engineers, Inc., www.ashrae.org.

Mermelstein (1998) describes a two-stage freezer unit capable of handling up to $2750\,kg\,h^{-1}$ of seafood pieces. Poulsen (1986) compared a number of freezer types in terms of refrigeration load and power consumption for the same food inlet temperature (10°C) and refrigeration evaporating temperature (−43°C), albeit for very different food products. He quotes an electrical power consumption of 120 kW for a load of $523\,MJ\,h^{-1}$ (i.e. 145 kW) when freezing peas. No further data are given but by assuming mean heat capacities in the unfrozen and frozen states of $3.3\,kJ\,kg^{-1}\,K^{-1}$ and $1.8\,kJ\,kg^{-1}\,K^{-1}$ respectively, a latent heat of $250\,kJ\,kg^{-1}$, a freezing temperature of −2°C and a final product temperature of −30°C, the heat load, calculated from equation 3.5, is $340\,kJ\,kg^{-1}$ which together with the quoted heat load suggests a throughput of $1535\,kg\,h^{-1}$.

Data comparing the operation of fluidized beds with other types of freezing plant are also scarce. The models developed by Reynoso and Calvelo (1985), which are based on Plank's equation for the prediction of freezing time, compare the production rate of peas in a fluidized bed freezer with that in a fixed bed continuous freezer of the same dimensions. At air velocities of 1.2 u_{mf} and 0.8 u_{mf} in the fluidized and fixed beds respectively, the production rate in the fluidized bed freezer was 20–30% greater whilst the power consumption was 1.7–2.1 times greater for the fixed bed. It is suggested that the explanation lies in the lack of vertical mixing in a fixed bed and that freezing time is controlled by particles in the top layer of the bed which are at a higher temperature. The increased capacity of a fluidized bed is attributed largely to the existence of particle mixing rather than the marginally higher heat transfer coefficient.

Freezing rate and freezing point of foods

Figure 3.3 shows a typical freezing curve which results when heat is removed at a continuous rate from foodstuffs and the temperature decreases. The curve has three zones: first, the removal of sensible heat from the food between the initial temperature and the freezing temperature; second, the removal of the latent heat of fusion leading to a change of state and the formation of ice crystals; and third, further sensible heat removal down to the required storage temperature. A number of features of the freezing curve require explanation. Whilst

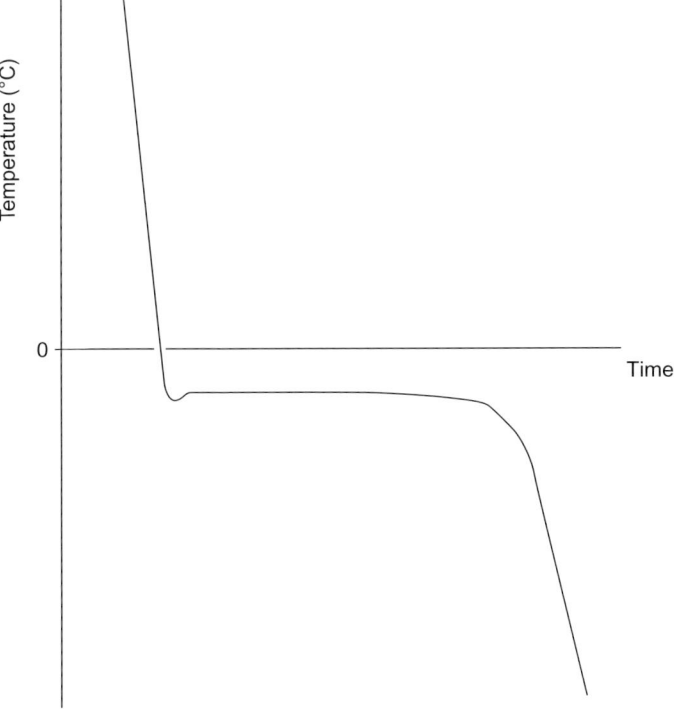

Figure 3.3　Typical food freezing curve.

pure water has a distinct freezing point (0°C), foods do not have clear freezing points. Rather, there is a freezing range which is normally between 0°C and about –2 or –3°C. It is perhaps more appropriate to refer to a freezing temperature (rather than freezing point) which is sometimes taken to be an average temperature across the broadly constant temperature horizontal portion of the curve (Pham, 1986). If a freezing point is to be specified then it is usually defined as the highest temperature at which ice crystals are found to be stable.

Immediately before the formation of ice crystals there is a degree of supercooling which initiates the formation of ice nuclei. This is apparent as the slight dip in temperature before the horizontal portion of the curve. During the removal of latent heat the temperature remains approximately constant and would be truly constant during the freezing of pure water. However, the presence of soluble components in the food results in a freezing point depression; this effect is more pronounced with lower molecular weight solutes. Finally, as the water in the food forms ice crystals, the remaining liquid water has an increased concentration of these solutes and the freezing point is further depressed, resulting in a gradual curve on the temperature-time plot and in a freezing temperature range rather than a clear freezing point. Because significant amounts of liquid water may be present even at temperatures well below the nominal freezing point, there is not necessarily a clearly defined end to the freezing period. The freezing time, sometimes referred to as the thermal arrest time, is usually interpreted as the time to remove the latent heat of fusion from the food and corresponds to the broadly horizontal portion of the freezing curve.

The quality of frozen food is affected very significantly by the rate at which freezing takes place; that is, the rate at which heat is removed from the product, and the consequent ice crystal size (Arthey, 1993). Slow freezing produces a small number of larger nuclei which then grow into large ice crystals. These tend to grow in the spaces between cells and result in structural damage to cell walls. In vegetable matter the disruption of cellular integrity leads to a loss of rigidity in cell walls and an inability of the food to retain water. On thawing, a soft, flaccid texture will result when the vegetable is cooked. In flesh foods the elasticity of the cellular structure minimises the disruptive effect of ice crystal formation and the loss of quality is associated largely with a loss of protein functionality; the associated loss of texture and water-holding capacity again results in a poor-quality thawed product. Rapid freezing, on the other hand, results in a high nucleation rate and the subsequent growth of a large number of small ice crystals which grow both within and outside cells. As a result the cells maintain their integrity which in turn minimises drip loss during thawing. Holdsworth (1983) suggests that the effect of freezing rate on quality is rather more significant with high moisture content foods, for example strawberries and tomatoes, than with high starch content materials.

Holdsworth (1983) defines slow freezing, such as might be achieved in a bulk freezing room, as a linear freezing rate of 0.2 cm h^{-1} (5.6 × 10^{-7} m s^{-1}). Conventional air blast or plate freezers result in quick freezing in the range 0.4–3.0 cm h^{-1} (1.1 × 10^{-6} to 8.3 × 10^{-6} m s^{-1}), whilst fluidized bed freezing is described as rapid with freezing rates as high as 5–10 cm h^{-1} (1.4 × 10^{-5} to 2.8 × 10^{-5} m s^{-1}). Marin *et al.* (1983) report experimental freezing rates of up to 11.8 cm h^{-1} (3.27 × 10^{-5} m s^{-1}) for potato spheres of 20 mm diameter fluidized in inert glass beads; freezing rates decreased markedly with diameter. These authors conclude that this technique offers economic benefits compared to air blast or cryogenic freezing if the product thickness is below 30 mm.

Prediction of freezing time

With respect to an individual food piece, the unit operation of freezing involves unsteady-state heat transfer; in other words, the temperature of the food changes with time. In these circumstances heat transfer by conduction is described by Fourier's first law

$$\frac{\partial T}{\partial t} = \alpha \left(\frac{\partial^2 T}{\partial x^2} + \frac{\partial^2 T}{\partial y^2} + \frac{\partial^2 T}{\partial z^2} \right) \qquad 3.1$$

which relates the temperature variation in three orthogonal dimensions x, y and z to time t. The thermal diffusivity α contains the physical properties of the material through which heat is transferred and is defined by

$$\alpha = \frac{k}{\rho c_p} \qquad 3.2$$

where k is the thermal conductivity, ρ is the density and c_p is the heat capacity. Thermal diffusivities are of the order of 10^{-7} m^2 s^{-1} for most foods.

Fourier's equation does not take into account convection, which in the case of food freezing governs heat transfer between the food surface and the refrigerating medium, for example cold air in the case of a fluidized bed. Second, and more importantly, it cannot account for the removal of latent heat and the resultant phase change. Of the models available to predict freezing time, Plank's equation (Plank, 1913, 1941) is one of the simplest and most widely used and is derived in many standard texts (Singh and Heldman, 2001; Smith, 2003). The principal assumption is that heat transfer is sufficiently slow to approximate to steady state. Thus for a foodstuff of density ρ, thermal conductivity k, freezing temperature T_f and latent heat of fusion λ, the freezing time according to Plank is

$$t = \frac{\rho\lambda}{(T_f - T_1)}\left[\frac{Pa}{h} + \frac{Ra^2}{k}\right] \qquad 3.3$$

where T_1 is the temperature of the surrounding medium. The relevant dimension a of the food for different geometries, together with the values of the parameters P and R, are summarised in Table 3.2. Of the bodies in Table 3.2, only the sphere is finite. Plank's model can be extended to predict the freezing time of cuboids (Smith, 2003) whilst the model proposed by Pham (1986) can be extended to cuboids and finite cylinders (Singh and Heldman, 2001). There is no clear definition of the latent heat of fusion for food materials. It is usually assumed that the latent heat is equal to the product of the latent heat of water and the moisture content of the food. An alternative is to use the change in enthalpy between the freezing temperature and an arbitrary temperature at which freezing is considered to be complete. Rao and Rizvi (1995) give comprehensive tables of the enthalpies of various foods as a function of temperature.

Nagaoka's equation (Nagaoka *et al.*, 1955) is an extension of Plank's model and takes into account the time required to reduce the temperature from an initial temperature T_i above the freezing point. The latent heat of fusion in equation 3.3 is replaced by the total enthalpy change Δh which includes the sensible heat which must be removed in reducing the temperature from an initial T_i, and in addition an empirical correction factor is included. Thus

$$t = \frac{\rho\Delta h}{(T_f - T_1)}[1 + 0.008(T_i - T_f)]\left[\frac{Pa}{h} + \frac{Ra^2}{k}\right] \qquad 3.4$$

Table 3.2 Values of parameters in Plank's equation.

	P	R	a
Infinite slab	$\dfrac{1}{2}$	$\dfrac{1}{8}$	Thickness*
Infinite cylinder	$\dfrac{1}{4}$	$\dfrac{1}{16}$	Diameter
Sphere	$\dfrac{1}{6}$	$\dfrac{1}{24}$	Diameter

* When frozen from both surfaces; a = half thickness for freezing from one surface.

where the total enthalpy change Δh, or heat load, during the freezing process is given by

$$\Delta h = c_{p_u}(T_i - T_f) + \lambda + c_{p_f}(T_f - T_{final}) \qquad 3.5$$

Here T_{final} is the final temperature and c_{p_u} and c_{p_f} are the heat capacities of the unfrozen and frozen food respectively.

The use of the Plank/Nagaoka model can be illustrated with a simple example. Consider a fluidized bed 0.75 m wide and 5 m long which is used to freeze peas 8 mm in diameter at a rate of 6000 kg h^{-1}. Assume that the peas enter the bed at 12°C, have a freezing temperature of −2°C and that the fluidizing air enters the bed at −35°C and at a velocity such that the heat transfer coefficient (see Heat transfer in fluidized bed freezers, below) is 170 W m^{-2}K^{-1}. What is the necessary bed depth?

Taking the density, thermal conductivity and latent heat of fusion of the peas to be 1050 kg m^{-3}, 1.0 W m^{-1}K^{-1} and 250 kJ kg^{-1} respectively and the bulk density within the fluidized bed as 525 kg m^{-3}, the Plank/Nagaoka model can be used to estimate the time required for the product outlet temperature to reach −20°C.

A simple enthalpy balance gives the total heat load per unit mass as

$$\Delta h = 3.3(12 - (-2)) + 250 + 1.8(-2 - (-20)) \text{ kJ kg}^{-1}$$

if the heat capacities of the unfrozen and frozen peas are assumed to be 3.3 and 1.8 kJ kg^{-1}K^{-1} respectively. Thus

$$\Delta h = 328.6 \text{ kJ kg}^{-1}$$

The freezing time using Nagaoka's equation is then

$$t = \frac{1050 \times 328.6 \times 10^3}{(-2 - (-35))}[1 + 0.008(12 - (-2))]\left[\frac{8 \times 10^{-3}}{6 \times 170} + \frac{(8 \times 10^{-3})^2}{24 \times 1.0}\right] \text{s}$$

and

$$t = 122 \text{ s}$$

The freezing time must now be equal to the residence time in the bed (see Mixing, dispersion and residence time, below); a mean residence time can be assumed to be equal to the mass hold-up in the bed divided by the mass flow rate. If the mass hold-up is the product of bed volume and the bulk density of the bed, and the bed depth is H, then

$$H = \frac{6000 \times 122}{3600 \times 0.75 \times 5 \times 525}\,\text{m}$$

and

$$H = 0.103\,\text{m}$$

Design of fluidized bed freezers

Introduction

According to De Michelis and Calvelo (1994), the design of a fluidized bed freezer requires a knowledge of the gas-particle heat transfer coefficient, prediction of the freezing time (for which all the available models are essentially modifications of Plank's equation) and finally an axial dispersion model to determine the residence time in the bed. Ditchev and Richardson (1999) stress both the importance of the heat transfer coefficient between food and the freezing medium in the intensification of the freezing process, and therefore in increasing food quality, and the importance of being able to predict freezing time which in turn determines the residence time and hence the throughput.

Persson (1967) indicated that it is convenient to use the Froude number, defined by equation 1.2, to assess the 'degree of fluidization' and for peas suggested values of Fr between 40 and 45.

$$Fr = \frac{u_{mf}^2}{gd} \qquad\qquad 1.2$$

Using the Ergun equation to estimate minimum fluidizing velocity, this would give superficial velocities of approximately $2\,\text{m\,s}^{-1}$ for a diameter of 9 mm. However, the practical lower limit to Fr is higher than this, for example because of icing problems when handling wet material, and thus the lower limit is about $Fr = 65$ (Persson, 1967) and is a function of the product to be frozen. Persson (1967) summarised this information graphically (Figure 3.4), originally in Imperial units, and showed the limits of the food freezing zone as a function of Froude number, Reynolds number, particle size and superficial air velocity. Persson concluded that under normal conditions a fluidized bed freezer should operate at a Froude number of approximately 100–120.

Heat transfer in fluidized bed freezers

The gas velocity in a fluidized bed freezer is fixed by the particle size and by the requirements of fluidization and cannot be varied to increase

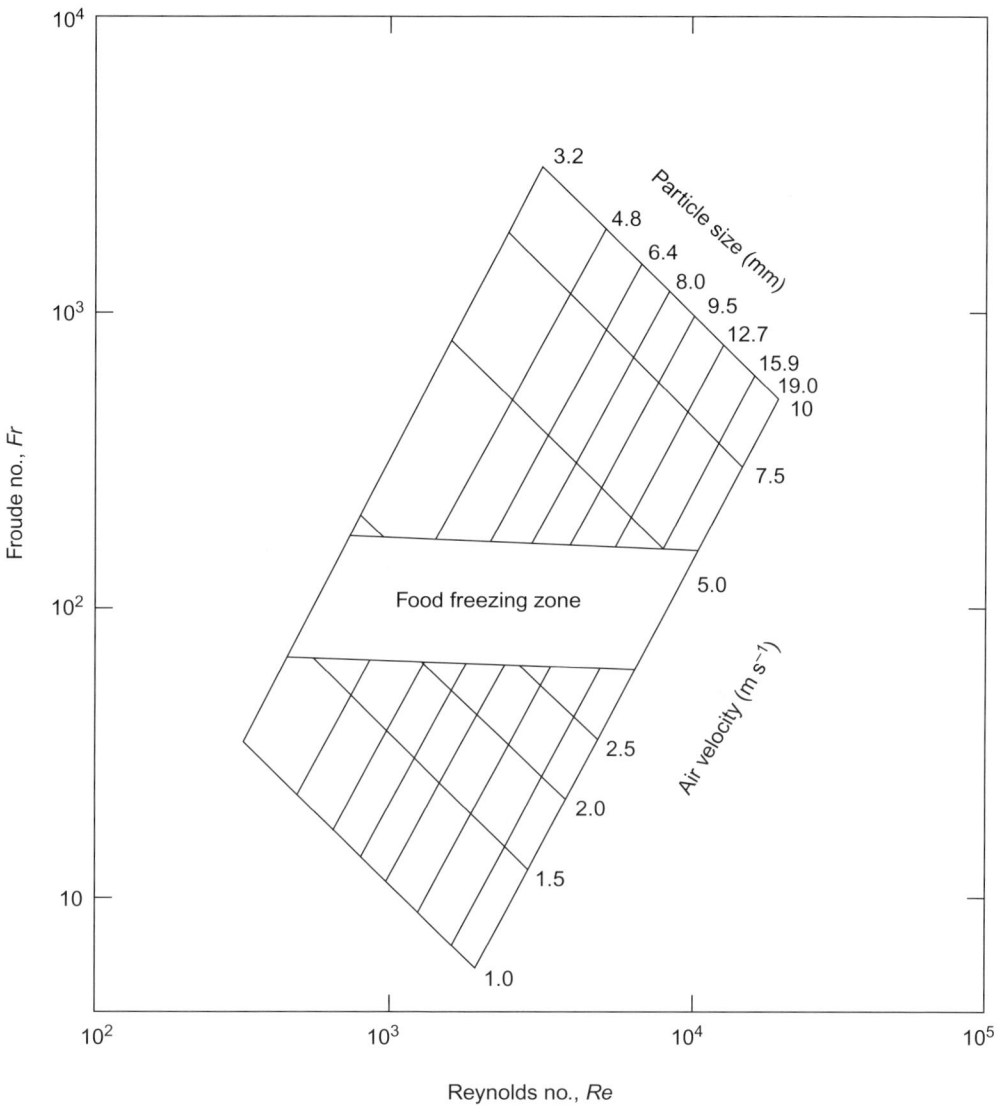

Figure 3.4   Limits of the food freezing zone as a function of Froude number, Reynolds number, particle size and superficial air velocity. Adapted from Persson, ASHRAE Journal, June 1967. © American Society of Heating, Refrigerating and Air-Conditioning Engineers, Inc., www.ashrae.org.

the rate of heat transfer, with the consequence that the rate of heat transfer can be increased only by lowering the temperature of the inlet gas (Holdsworth, 1987). A knowledge of the heat transfer coefficient is essential to the design of a freezer. The short freezing times for peas (see Applications of fluidized bed freezing, below), for example, indicate high heat transfer coefficients which are due partly to the high surface area available and partly to the instability of the laminar boundary layer brought about by particle motion (Holdsworth, 1987). Sheen and Whitney (1990) point out that an increase in the heat transfer coefficient reduces the freezing time but that there is a limit to this effect as the conductive thermal resistance of the food becomes the limiting factor. Jowitt (1977) suggested that data in Botterill (1975), who makes no direct reference to foods, give values of the gas-particle heat transfer coefficient in the range 64–1690 W m^{-2}K^{-1} for conditions corresponding to the fluidization of peas. There are very few studies which have attempted to determine heat transfer coefficients at conditions relevant to the freezing of food, i.e. in shallow beds at Reynolds numbers greater than 1000 (Vazquez and Calvelo, 1980). However, the available information suggests that the values quoted by Jowitt are excessive.

Despite this, the expected heat transfer coefficients obtainable in a fluidized bed are greater than those for forced convection in a gas (Ditchev and Richardson, 1999) although not as high as in the dynamic dispersion medium (DDM) method described by these authors. Comparative data are presented in Table 3.3.

For heat transfer to a single sphere in turbulent flow, McAdams (1954) suggested

$$Nu = 0.37 Re^{0.6} Pr^{0.33} \qquad \qquad 3.6$$

over the range $17 < Re < 7 \times 10^4$. However, a stronger dependence on Reynolds number was proposed in the frequently quoted correlation for the gas-particle heat transfer coefficient in fluidized beds due to Kunii and Levenspiel (1991)

$$Nu = 0.03 Re^{1.3} \qquad \qquad 3.7$$

Equation 3.7 is valid only for particle Reynolds numbers below 100 and the Reynolds number of, say, a pea in a fluidized bed freezer is of the order of 1000 and thus the correlation tends to overestimate the heat transfer coefficient considerably. Kelly (1965) proposed, specifically for the fluidization of peas,

Table 3.3 Comparison of heat transfer coefficients in fluidization.

	Superficial gas velocity (m s^{-1})	h (W m^{-2}K^{-1})
Free convection	–	8–15
Forced convection	2–5	10–50
Fluidization	2–5	60–120
DDM	0.5–2	180–420

$$Nu = 3.5 \times 10^{-4} Re^{1.5} \qquad\qquad 3.8$$

with a not dissimilar exponent on Re to that of Kunii and Levenspiel, although this expression, valid for $1000 < Re < 3000$, indicates far more realistic values of the heat transfer coefficient. Botterill (1975), commenting on the Kunii and Levenspiel correlation, suggested that the high dependence on Re is not consistent with other work and cited the work of Rabinovich *et al.* (1967), who proposed Re to the power 0.77 as being more reasonable. Values of the heat transfer coefficient calculated from the correlations due to Kunii and Levenspiel and to Kelly are compared in Table 3.4.

Persson (1967) measured the variation in gas-particle heat transfer coefficient with particle diameter. Figure 3.5 shows the heat transfer coefficient for both an unspecified 'maximum' gas velocity and for $Fr = 120$. For $Fr = 120$, h falls from approximately $130 \, \mathrm{W \, m^{-2} K^{-1}}$ at a diameter of $3 \, \mathrm{mm}$ to $100 \, \mathrm{W \, m^{-2} K^{-1}}$ at a diameter of $16 \, \mathrm{mm}$, these particle sizes corresponding to superficial gas velocities of $1.88 \, \mathrm{m \, s^{-1}}$ and $4.33 \, \mathrm{m \, s^{-1}}$ respectively. For larger diameters the heat transfer coefficient becomes approximately constant at $85 \, \mathrm{W \, m^{-2} K^{-1}}$.

There are considerable problems in measuring heat transfer coefficients where only sensible heat is transferred because of the difficulties in measuring particle surface temperatures in a fluidized bed (Vazquez

Table 3.4 Heat transfer coefficient as a function of particle diameter.

	h ($\mathrm{W \, m^{-2} K^{-1}}$)							
	$\dfrac{u}{u_{mf}} = 1$				$\dfrac{u}{u_{mf}} = 2$			
	Particle diameter (mm)				Particle diameter (mm)			
	1	5	10	20	1	5	10	20
Kunii and Levenspiel (1991) $Nu = 0.03 \, Re^{1.3}$ $Re < 100$	157.7	–	–	–	–	–	–	–
Kelly (1965) $Nu = 3.5 \times 10^{-4} \, Re^{1.5}$ $1000 < Re < 3000$	–	–	76.5	–	–	91.0	–	–
Vazquez and Calvelo (1980) $j_H = 0.097 \, Re^{-0.502} \, Ga^{0.195}$ $1400 < Re < 4000$	–	–	284.6	–	–	319.3	–	–
Vazquez and Calvelo (1983) $j_H = 0.204 \, Re^{-0.563} \, Ga^{0.179}$ $3300 < Re < 14000$	–	–	–	320.3	–	–	379.5	433.6

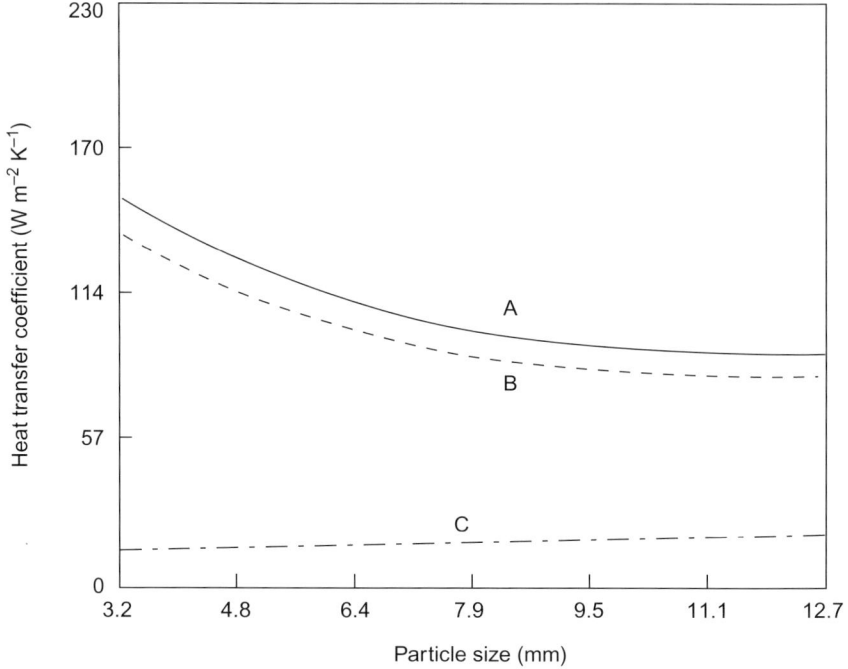

Figure 3.5 Variation in gas-particle heat transfer coefficient with particle diameter (A = fluidized bed at maximum air velocity; B = fluidized bed at Fr = 120; C = flat surface velocity as for curve B). Adapted from Persson, ASHRAE Journal, June 1967. © American Society of Heating, Refrigerating and Air-Conditioning Engineers, Inc., www.ashrae.org.

and Calvelo, 1980). These authors measured the gas-particle heat transfer coefficient h in a bed of drying particles in which the transfer of only latent heat was involved and thus the particles remained at the wet bulb temperature as long as drying was within the constant rate period. Vazquez and Calvelo (1980) outlined a model for the heat transfer coefficient based upon an energy balance across a differential element of the fluidized bed of height dH. Using the interstitial gas velocity u' and the bed voidage ε, the energy balance yields

$$dQ = (\rho u' \varepsilon)Ac_p dT = haA(T_S - T)dH \qquad 3.9$$

where Q is the rate of heat transfer, the bracketed term represents the superficial mass air flux G, i.e. the mass flow rate of air through the bed per unit bed cross-section, T is the air temperature and T_S is the particle surface temperature. Assuming that the particle surface area per unit bed volume a can be replaced for spherical particles of diameter d by

$$a = \frac{6(1-\varepsilon)}{d} \qquad 3.10$$

equation 3.9 becomes

$$dQ = Gc_p dT = \frac{6h(1-\varepsilon)(T_S - T)}{d} dH \qquad 3.11$$

which on integration across a bed height H and between inlet and outlet bed temperatures T_1 and T_2 respectively yields

$$\ln\left[\frac{T_S - T_2}{T_S - T_1}\right] = \frac{6h(1-\varepsilon)H}{Gc_p d} \qquad 3.12$$

Vazquez and Calvelo (1980) used peas 7 mm in diameter and alumina particles 5.6 mm in diameter with bed heights in the range 2.5–6 cm; experimental temperatures varied between 14°C and 23°C, obviously somewhat higher than the air temperatures used in fluidized bed freezing. For peas the gas-particle heat transfer coefficient was found to be independent of bed height and varied in the range 180–221 W m^{-2}K^{-1}. Good agreement was reported when these experimental data were plotted according to the Chilton-Colburn j-factor model (see, for example, Bird *et al.*, 1960; Kay and Nedderman, 1985). Thus

$$j_H = 3.375 Re^{-0.54} \qquad 3.13$$

where

$$j_H = Nu Re^{-1} Pr^{-\frac{1}{3}} \qquad 3.14$$

and the Reynolds number is defined by

$$Re = \frac{\rho u d}{\mu(1-\varepsilon)} \qquad 3.15$$

In order to take account of the effect of temperature and to determine the heat transfer coefficient at realistic freezing temperatures, rather than the room temperature of the reported experiments, and to extend the model to other particle shapes and sizes, Vazquez and Calvelo (1980) then plotted their data according to the model proposed by Chang and Wen (1966)

$$j_H = 0.097 Re^{-0.502} Ga^{0.195} \qquad 3.16$$

where Ga is the Galileo number (identical to the Archimedes number Ar) and is defined by equation 1.49. This equation correlates the data of Chang and Wen (1986), the data of Bradshaw and Myers (1963) and that of Vazquez and Calvelo (1980) who thus estimated that the gas-particle heat transfer coefficient increased by 13% as the air temperature decreased from 20°C to −40°C. The approximate experimental range of Reynolds number in their work was $1400 < Re < 4000$.

In an extension of their previous work, Vazquez and Calvelo (1983a) measured the heat transfer coefficient within beds of 0.89 cm potato cubes and potato chips (i.e. cuboids 0.92 cm by 0.92 cm by 5.3 cm). The effective diameter for these non-spherical particles was calculated from

$$d = d_e \phi$$ 3.17

where the equivalent spherical diameter of a particle of volume V is given by

$$d_e = \left(\frac{6V}{\pi}\right)^{\frac{1}{3}}$$ 3.18

and the sphericity by

$$\phi = \frac{\pi d^2}{A_p}$$ 3.19

A_p is the particle surface area and therefore the effective diameter becomes

$$d = \frac{6V}{A_p}$$ 3.20

The heat transfer coefficient for potato cubes was smaller than for fluidized beds of peas (Vazquez and Calvelo, 1980) and varied between 163 and 189 W m⁻²K⁻¹, over the superficial velocity range 2.5–4.3 m s⁻¹ (where u_{mf} = 1.57 m s⁻¹); in other words, it is approximately constant. Again, in these experiments, the temperature was unrealistically high and varied from 15°C to 23°C. The corresponding data for potato chips are: h = 160–175 W m⁻²K⁻¹ in the velocity range 2.8–4.3 m s⁻¹ (u_{mf} = 2.36 m s⁻¹), corresponding to Reynolds numbers in the range $3300 < Re < 14\,000$. The heat transfer coefficient was correlated, as in the earlier work, using the Chilton-Colburn j-factor model

$$j_H = 0.204 Re^{-0.563} Ga^{0.179}$$ 3.21

Figure 3.6 shows a comparison of the experimental data and equation 3.21. Confusingly, Vazquez and Calvelo (1983a) also appear to

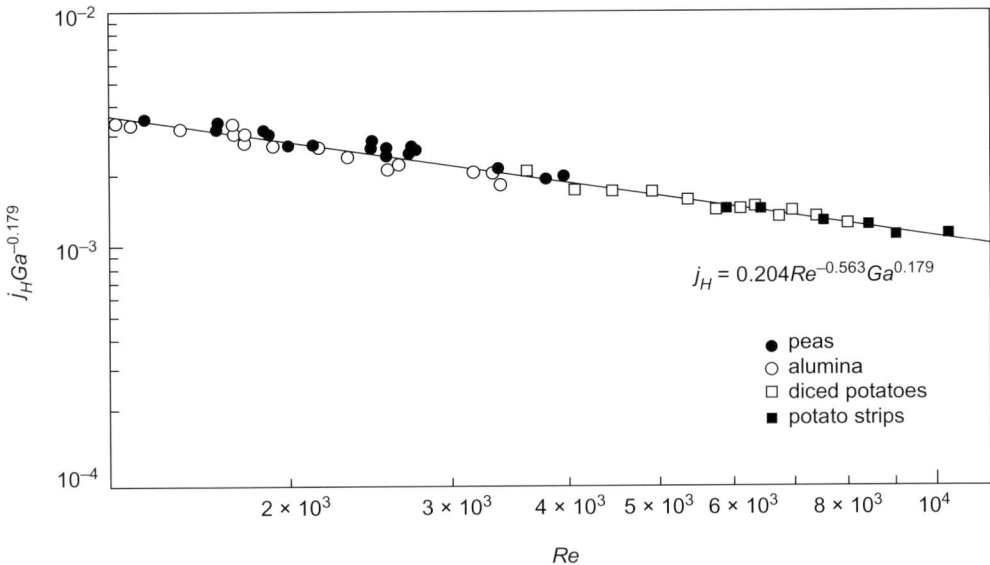

Figure 3.6   Experimental heat transfer coefficients in fluidized bed freezing. From Vazquez and Calvelo (1983a), by permission of the Institute of Food Technologists, USA.

claim that this relationship represents the data for peas presented in their earlier paper.

Heat transfer coefficients determined from the correlations proposed by Kunii and Levenspiel (1991), Kelly (1965) and Vazquez and Calvelo (1980, 1983a) (equations 3.7, 3.8, 3.16 and 3.21 respectively) are compared in Table 3.4 and in Figure 3.7. For this comparison the heat transfer coefficient has been calculated for a range of particle sizes based upon, first, a particle density of $1050\,kg\,m^{-3}$ and a voidage at minimum fluidizing conditions of 0.5 and, second, an air temperature of $-35°C$ at which the density, viscosity, heat capacity and thermal conductivity of air are $1.5\,kg\,m^{-3}$, $1.5 \times 10^{-5}\,Pa\,s$, $1.0\,kJ\,kg^{-1}\,K^{-1}$ and $0.021\,W\,m^{-1}\,K^{-1}$ respectively. The simplified Ergun equation (equation 1.50) was used to estimate minimum fluidizing velocity. Table 3.4 lists the values of heat transfer coefficient calculated from each of the listed correlations, for particle diameters of 1, 5, 10 and 20 mm, which fall within the respective range of Reynolds number. Figure 3.7 shows the variation of heat transfer coefficient over the full range of valid particle diameter (assuming the conditions set out above) for each correlation. Kunii and Levenspiel (1991) grossly overestimate the heat transfer coefficient at Reynolds numbers greater than 100 and therefore their model can be discounted for the food particles likely to be found in food freezing operations. Accordingly, these data have been omitted from Figure 3.7.

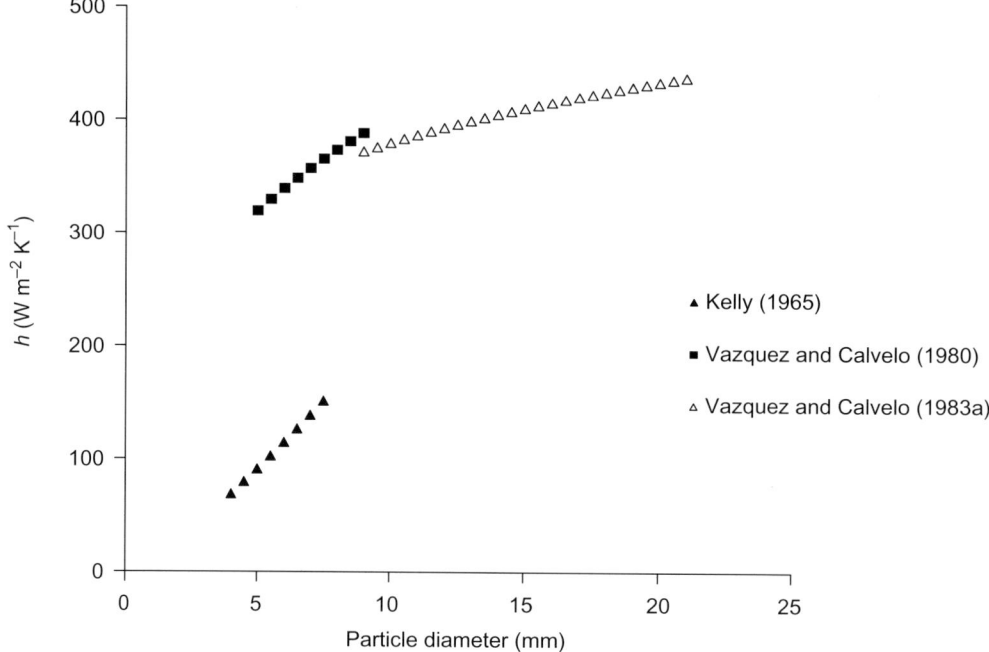

Figure 3.7   Predicted heat transfer coefficient as a function of particle size.

The correlation due to Kelly (1965) indicates relatively modest coefficients below $200\,\mathrm{W\,m^{-2}\,K^{-1}}$ for particles up to about 8mm in diameter. The equations proposed by Vazquez and Calvelo, based on the Chilton-Colburn j-factor model, suggest rather higher coefficients of the order of 300–$400\,\mathrm{W\,m^{-2}\,K^{-1}}$ for the fluidization of particles up to 20mm in diameter at gas velocities reasonably in excess of those required for minimum fluidization.

The heat transfer coefficient in the Plank/Nagaoka model (equation 3.4) is defined by

$$q = h(T_S - T_1) \qquad\qquad 3.22$$

where T_S is the particle surface temperature and T_1 is the temperature of the refrigerated air. However, Vazquez and Calvelo (1983b) point out that the temperature of the air changes as it passes through the bed (see Figure 3.8) and that the outlet temperature T_{g2} varies with bed height. Instead, an effective heat transfer coefficient h_e should be used as defined by equation 3.23

$$q = h_e(T_S - T_{g1}) \qquad\qquad 3.23$$

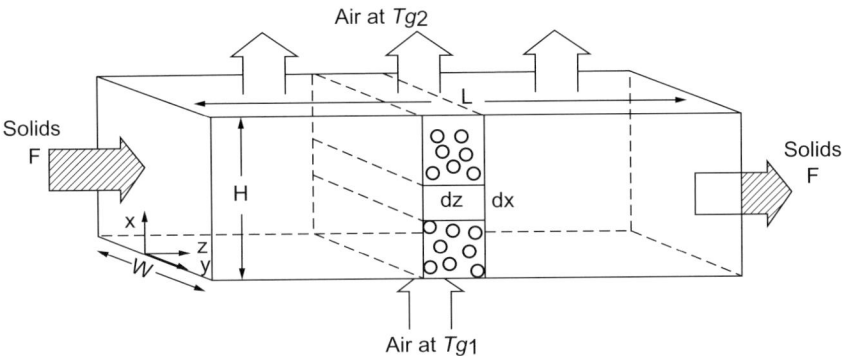

Figure 3.8 Variation of air temperature with bed height. From Vazquez and Calvelo (1983b), by permission of the Institute of Food Technologists, USA.

where T_{g_1} is the bed inlet temperature. Further, these authors show that h_e can be obtained from an energy balance over a differential bed length giving

$$h_e = \frac{Gc_p}{aH}\left[1 - \exp\left(\frac{-haH}{Gc_p}\right)\right] \qquad 3.24$$

Thus the value of h_e approaches h as the bed height H approaches zero.

Rather more information appears to be available about heat transfer coefficients in the so-called DDM method of freezing. Marin *et al.* (1983) measured the heat transfer coefficient for potato spheres between 20 and 40 mm in diameter, either fixed in the bed or freely floating, when immersed in a bed of glass beads (between 315 and 400 μm in diameter) fluidized at velocities 2–3 times greater than the minimum. Coefficients are approximately 230 W m^{-2} K^{-1} compared to only 35 W m^{-2} K^{-1} for the same bodies suspended above the bed surface, i.e. equivalent to air blast freezing. They concluded that fluidization allows rapid freezing (with $h = 200$ W m^{-2} K^{-1}) below a product thickness of 30 mm and contrasted fluidized beds with cryogenic processes; the latter give slightly more rapid freezing ($h > 500$ W m^{-2} K^{-1}) than fluidization but the associated operating costs are considerable. These findings are summarised in Figure 3.9. In addition, this same group of workers (Rios *et al.*, 1984) suggest heat transfer coefficients as high as 250 W m^{-2} K^{-1} for either fixed or freely moving food pieces *larger* than 30 mm diameter when placed in a bed of 400 μm inert particles of density 2600 kg m^{-3}.

More modest surface-to-bed heat transfer coefficients have been reported for a DDM fluidized bed, albeit during heating of the bed, using large epoxy resin-coated particles, either 8 mm or 12 mm in

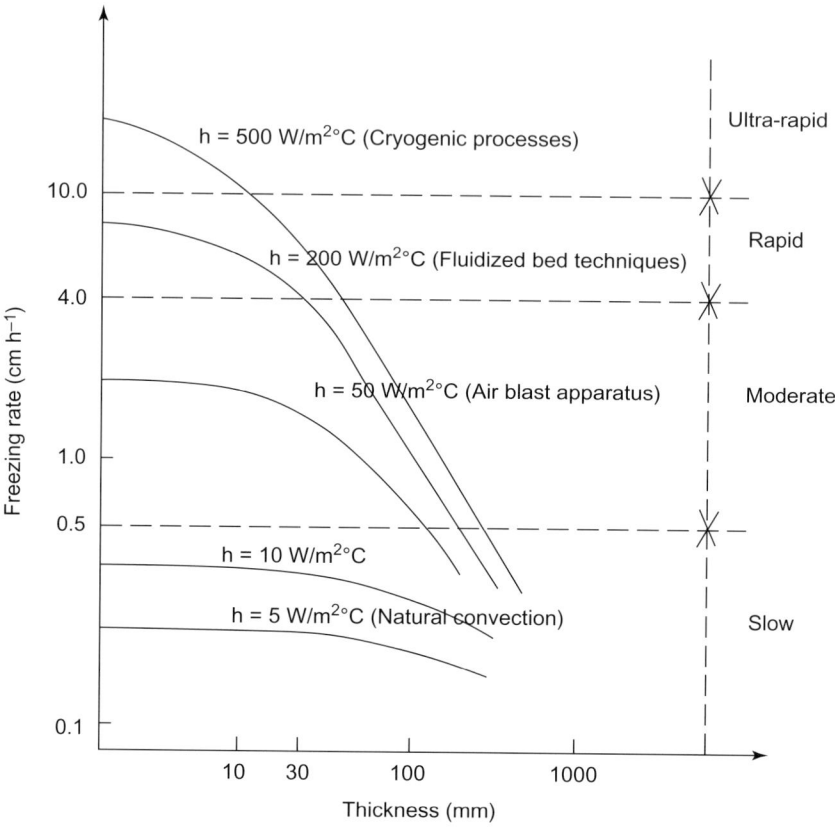

Figure 3.9 Influence of product thickness and heat transfer coefficient on mean freezing rate for potato slices. From Rios, G.M., Marin, M. and Gibert, H., New developments of fluidization in the IQF food area, in: McKenna, B.M., (ed.), Engineering and food, volume 2, processing applications, International Congress on Engineering and Food, 1983, Elsevier, 1984, figure 3. With kind permission of Springer Science and Business Media.

diameter, and particle density of $510 \, \text{kg} \, \text{m}^{-3}$ (Sheen and Whitney, 1990). These authors used an electrical resistance heating element within a copper cylinder, 28.6 mm in diameter and 165.1 mm in length, placed axially in the bed and reported surface-to-bed coefficients 2–3 times higher than those obtained with forced convection in air: 60–70 $\text{W} \, \text{m}^{-2} \text{K}^{-1}$ for 8 mm spheres and 80–90 $\text{W} \, \text{m}^{-2} \text{K}^{-1}$ for 12 mm spheres. Their experimental results could be predicted to within 5 % by

$$Nu = 37.47 \, Re^{0.222} \left(\frac{H_0}{L_H} \right)^{0.436} \left(\frac{d}{L_H} \right)^{-0.158} \qquad 3.25$$

where H_0 is the packed bed height and L_H is the vertical length of the heat transfer surface. When the heating element was replaced with a food material (carrot or sausage) of the same dimensions, the experimentally determined freezing time decreased in moving from free convection freezing ($h = 11.5\,\mathrm{W\,m^{-2}\,K^{-1}}$) to fluidized bed freezing where the upper limit to the heat transfer coefficient was 120–$150\,\mathrm{W\,m^{-2}\,K^{-1}}$.

Mixing, dispersion and residence time

Vazquez and Calvelo (1983b) presented a model for the prediction of the minimum residence time in a fluidized bed freezer which can then be equated to the required freezing time. The model is defined in terms of a longitudinal dispersion coefficient D_z which is a measure of the degree of solids mixing within the bed in the direction of flow (and has the dimensions of a diffusivity, and hence units of $\mathrm{m^2\,s^{-1}}$), a dimensionless time τ

$$\tau = \frac{V_z t}{L} \qquad\qquad 3.26$$

where V_z is the longitudinal velocity of the fluidized solids in the direction of flow, L is bed length and t is time, and the Peclet number Pe. The Peclet number is defined by

$$Pe = \frac{V_z L}{D_z} \qquad\qquad 3.27$$

However, substituting for the definition of V_z this becomes

$$Pe = \frac{FL}{\rho_s(1-\varepsilon)HWD_z} \qquad\qquad 3.28$$

where H and W are the bed height and width respectively, ρ_s is the particle density, and F is the mass flow rate of solids through the bed. Vazquez and Calvelo, using bed lengths up to 1.55 m, bed widths up to 0.40 m and 5.6 mm diameter peas fluidized at $1.19\,\mathrm{m\,s^{-1}}$, made a step change to coloured peas at the bed inlet and measured the mass fraction of these at the bed outlet every 30 seconds. They defined the minimum residence time t_{min} as the time required for a step change in the feed to decay to a mass fraction of 0.05 of coloured peas at the bed outlet. For high values of the Peclet number, i.e. $Pe > 2$, the solution of the material balance is

$$Pe = \frac{5.41\tau_{min}}{(1-\tau_{min})^2} \qquad\qquad 3.29$$

Values of the dispersion coefficient were obtained from the experimental values of t_{min} and correlated with bed height and superficial gas velocity as follows

$$D_z = 0.0951 H^{2.60} u^{3.54} \qquad 3.30$$

Good agreement is reported between predicted and experimental values of D_z, with values of the order of $10^{-4} \, m^2 s^{-1}$ over the velocity range $2.0–2.80 \, m s^{-1}$ and bed heights between 3.4 and 6.3 cm.

This work was extended by De Michelis and Calvelo (1994) who measured dispersion coefficients for 1 cm cubes and for 1 cm × 1 cm × 1.5 cm cuboids. The data were again correlated by an expression taking the form of equation 3.30, with similar exponents on bed height (2.46 and 2.58 for cubes and cuboids respectively) and gas velocity (3.13 and 3.34 respectively) but coefficients of 0.110 and 0.256 respectively; the coefficient in equation 3.30 increased with decreasing particle sphericity.

The strong dependence of D_z on bed height and gas velocity is explained by the mechanism of particle mixing in aggregative fluidization in which particle movement is due solely to the rise of bubbles, and the volumetric rate of particle movement depends upon bubble volume which in turn increases with bed height and gas velocity (see Chapter 1). The procedure is now to determine the dispersion coefficient for a given gas velocity from equation 3.30 and to determine t_{min} from equation 3.29 as a function of the dimensions of the bed, the feed flow rate, the physical properties of the particle bed and D_z. The minimum residence time is then equated to the required freezing time of a single particle.

Substituting equations 3.26 and 3.27 into equation 3.29 results in an expression for the mass production rate per unit plate area, thus

$$\frac{F}{WL} = \frac{H\rho_s(1-\varepsilon)}{t}\left[1 - \sqrt{\frac{5.41 t D_z}{L^2}}\right] \qquad 3.31$$

Vazquez and Calvelo (1983b) used this equation to determine an optimum bed height; Figure 3.10 shows production rate as a function of bed height for various bed lengths based on the freezing of peas down from 15°C to –18°C using air at an inlet temperature of –35°C and a gas velocity of $2.8 \, m s^{-1}$. This is explained by the interaction of three effects produced by an increase in bed height: an obvious increase in production rate; a decrease in the effective heat transfer coefficient leading to an increased freezing time; and an increase in the dispersion coefficient with a corresponding reduction in production rate (as predicted by equation 3.31). De Michelis and Calvelo (1985) used this

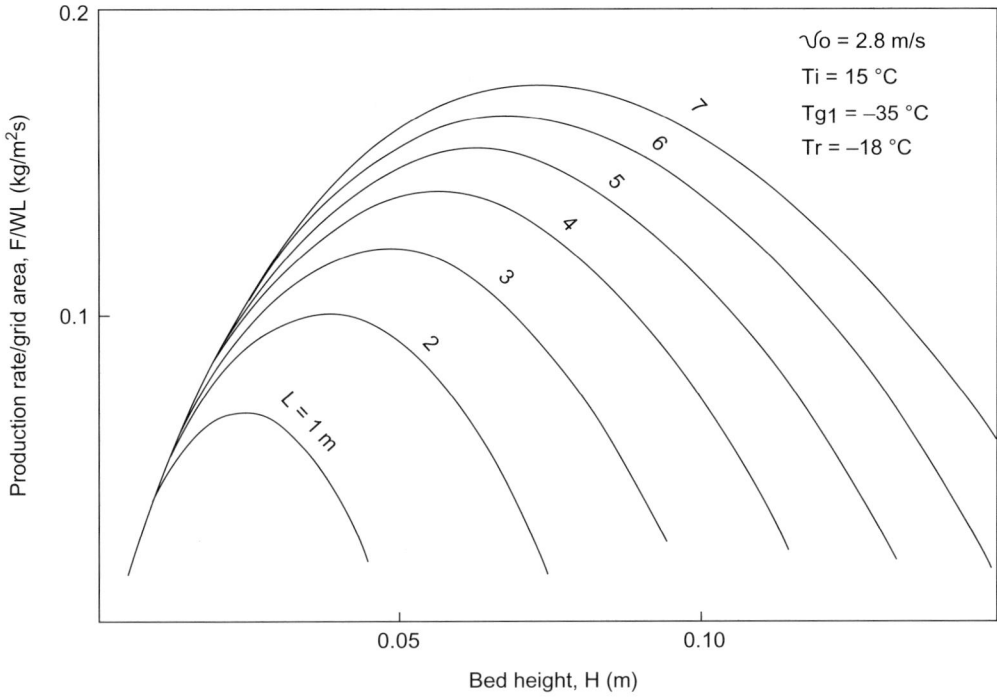

Figure 3.10 Production rate of frozen peas per unit plate area as a function of bed height and bed length. From Vazquez and Calvelo (1983b), by permission of the Institute of Food Technologists, USA.

model to optimise production rate in a fluidized bed freezer. Their results are presented in the form of a nomogram (Figure 3.11) for maximum production rate as a function of superficial velocity, bed height and air inlet temperature for a given particle size.

Applications of fluidized bed freezing

Peas are one of the largest volume products frozen using fluidized beds. In the UK a large processing capacity is needed for the very short pea season in midsummer; peas are then placed in long-term bulk storage for up to 12 months, followed by packaging at a later date. The freezing time for peas has been quoted variously as 110s to reach a temperature of −18°C using air at −30°C and a velocity of 3.7 m s^{-1}, reducing to 78s at a velocity of 6 m s^{-1} (Holdsworth, 1987); and approximately 3–4 minutes (Tressler *et al.*, 1968; Fennema *et al.*, 1973; Canet, 1989).

Following the commercial success of frozen peas, fluidization was subsequently extended to less ideally fluidizable foods such as potato

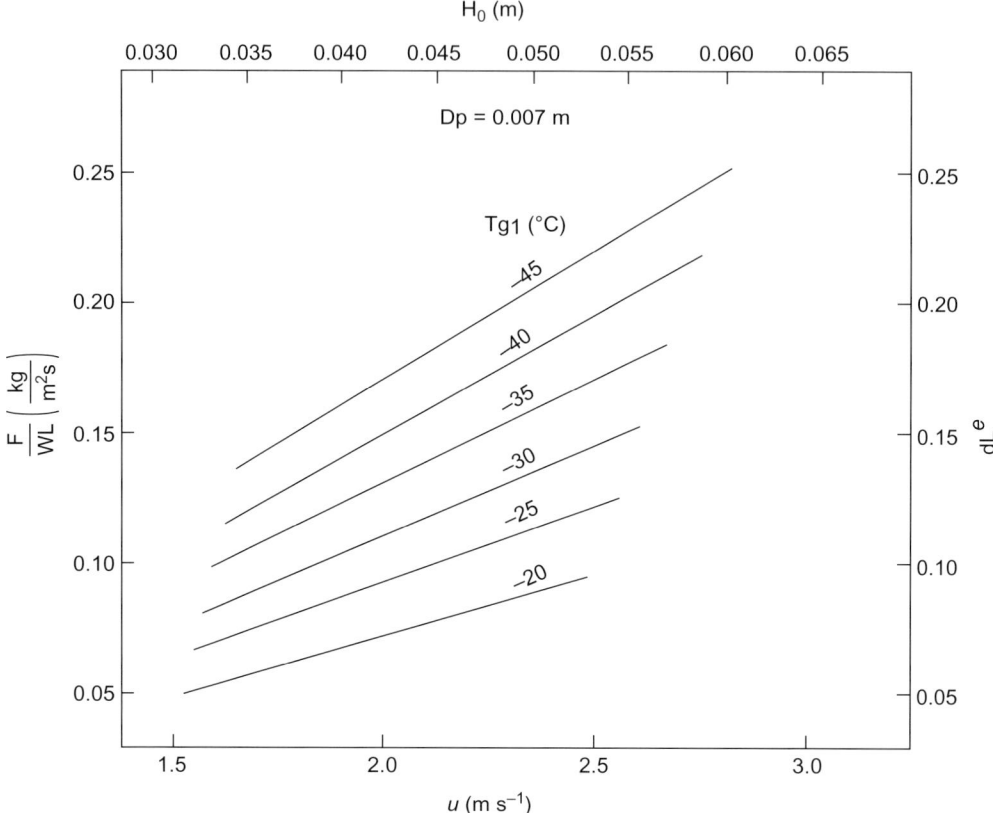

Figure 3.11   Nomogram for calculation of maximum production rate as a function of superficial velocity, bed height and air inlet temperature. From De Michelis and Calvelo (1985), by permission of the Institute of Food Technologists, USA.

chips, sprouts and green beans, often with assistance from vibrating conveyors (Jowitt, 1977). Indeed, the freezing of French fried potatoes is a major application of fluidized bed freezing (Persson and Londahl, 1993); potatoes are partially cooked and then cooled to 32°C before being frozen down to −23°C (Holdsworth, 1983). In addition, IQF products can be obtained from a wide range of vegetables which are liable to stick together such as green beans, sliced carrot and sliced cucumber (Persson and Londahl, 1993).

Soft fruits which are difficult to freeze in other ways represent the other large group of foods to be frozen in fluidized beds. Moist, heavy, tender and highly perishable fruits such as strawberries, raspberries and blackberries are the most difficult fruits to yield an IQF product. Fruit is pre-cooled to the freezing temperature in cooling rooms over a period of 8–10 hours in order to minimise weight loss and to shorten freezing times (Conroy and Ellis, 1981). Strawberries are cooled to −1°C and blackberries to −2.8°C, resulting in freezing times of eight minutes for strawberries and six minutes for blackberries. Canet (1989) suggests that the freezing time for strawberries is up to 15 minutes.

Fish and meat products are also frozen using fluidized bed freezers and Persson and Londahl (1993) suggest that this is the preferred

Table 3.5 Reported uses of fluidized bed freezing.

Vegetables	Corn	B
	Cut okra	L
	Diced blanched potatoes	L
	Diced carrots	E, L
	Diced squash	L
	Diced vegetables	B, K
	French fried potatoes	E, F, G, K, L
	Green beans	B, E, G, K, L
	Lima beans	B, E, L
	Onion rings	L
	Peas	A, B, F, E, L
	Sliced carrot	K
	Sliced cucumber	K
	Sprouts	B, G
	Whole kernel corn	E
Fruit	Apricot halves	L
	Blackberries	D
	Blueberries	B, E, L
	Cherries	L
	Diced pears	L
	Grapefruit sections	L
	Pineapple cubes	L
	Raspberries	D
	Seedless grapes	L
	Sliced apple	B, L
	Sliced peaches	L
	Sliced pineapple	B
	Strawberries	B, D, E, F, L
	Sweet melon cubes	L
Fish	Fish cakes	L
	Fish sticks	E, L
	Peeled cooked shrimps	K
	Raw fish portions	L
	Seafood pieces	H
	Shrimps	E, J
	Small lobster tails	L
	Whole small fish	L
Other	Bakery products containing yeasts	C
	Cut-up chicken	L
	Diced meat products	K
	Fried meatballs	K

Key to references
A: Arthey (1993); B: Canet (1989); C: Chen and Pei (1986); D: Conroy and Ellis (1981); E: Fennema *et al.* (1973); F: Holdsworth (1983); G: Jowitt (1977); H: Mermelstein (1998); J: Morrison (1993); K: Persson and Londahl (1993); L: Tressler *et al.* (1968)

Table 3.6   Reported freezing times in fluidized bed freezing.

	Freezing time (minutes)
Peas	3–4
Whole kernel corn	3–4
Lima beans	4–5
Blueberries	4–5
Diced carrots	6
Cut green beans	5–12
French fried potatoes	8–12
Strawberries	9–13
Small shrimps	6–8
Large shrimps	12–15
Fish sticks	15

method for fried meatballs, diced meat products and peeled cooked shrimps. Mermelstein (1998) reports a freezing time of two minutes for seafood pieces frozen using nitrogen vapour as the fluidizing gas. A rather different application is reported by Chen and Pei (1986) who proposed to freeze bakery products containing yeasts. Internal heat generation in the food may prolong freezing times in other types of freezer and thus adversely affect quality. The high heat transfer coefficient in fluidized beds is an alternative to using cryogenic methods in order to increase freezing rates. Carbon dioxide (i.e. dry ice) used as the fluidizing gas absorbs twice as much heat per unit mass as liquid nitrogen and therefore a faster freezing rate is obtained despite a lower temperature driving force and consequently operating costs are reduced. Other applications include packaged meat, fish fingers and shrimps using immersion freezing or the dynamic dispersion medium (DDM) method. The range of applications is summarised in Table 3.5 and some reported freezing times are listed in Table 3.6.

Nomenclature

a characteristic length; particle surface area per unit bed volume
A bed cross-sectional area
A_p particle surface area
Ar Archimedes number
c_p heat capacity at constant pressure
c_{p_f} heat capacity of frozen food
c_{p_u} heat capacity of unfrozen food
d particle diameter

d_e equivalent spherical diameter
D_z longitudinal dispersion coefficient
F mass flow rate of solids
Fr Froude number
g acceleration due to gravity
G superficial mass air flux
Ga Galileo number
h heat transfer coefficient; enthalpy
h_e effective heat transfer coefficient
H bed height
H_0 packed bed height
j_H j-factor for heat transfer
k thermal conductivity
L bed length
L_H vertical length of heat transfer surface
Nu Nusselt number
P parameter in Plank's equation
Pe Peclet number
Pr Prandtl number
q heat flux
Q rate of heat transfer
R parameter in Plank's equation
Re Reynolds number
t time; freezing time
t_{min} minimum residence time
T temperature
T_1 temperature of freezing medium; inlet bed temperature
T_2 outlet bed temperature
T_f freezing temperature
T_{final} final temperature of frozen food
T_{g_1} inlet bed temperature
T_{g_2} outlet bed temperature
T_i initial temperature
T_S particle surface temperature
u superficial velocity
u' interstitial gas velocity
u_{mf} minimum fluidizing velocity
V particle volume
V_z longitudinal velocity of fluidized solids
W bed width
x distance in the x dimension
y distance in the y dimension
z distance in the z dimension

Greek symbols

α	thermal diffusivity
Δh	total enthalpy change
ε	bed voidage
λ	latent heat of fusion
μ	viscosity
ρ	density
ρ_s	particle density
ϕ	sphericity
τ	dimensionless time

References

Arthey, D., Freezing of vegetables and fruits, in: Mallett, C.P., (ed.), Frozen food technology, Blackie, Glasgow, 1993.

Bird, R.B., Stewart, W.E. and Lightfoot, E.N., Transport phenomena, Wiley, New York, 1960.

Botterill, J.S.M., Fluid-bed heat transfer, Academic Press, London, 1975.

Bradshaw, R.D. and Myers, J.E., Heat transfer and mass transfer in fixed and fluidized beds of large particles, *A.I.Ch.E.J.*, **9** (1963) 590–595.

Canet, W., Quality and stability of frozen vegetables, in: Thorne, S., (ed.), Developments in food preservation, volume 5, Elsevier, London, 1989, 1–50.

Chang, T.M. and Wen, C.Y., Fluid to particle heat transfer in air fluidized beds, *Chem. Eng. Prog. Symp. Series*, **62** (67) (1966) 111–116.

Chen, P. and Pei, D.C.T., The freezing of bakery products, in: Le Maguer, M. and Jelen, P., (eds.), Food engineering and process applications, volume 2: unit operations. Fourth International Congress on Engineering and Food, Edmonton 1985, Elsevier, London, 1986, 79–87.

Conroy, R. and Ellis R.F., Pre-cooling berries improves quality, shortens fluidized freezing time, *Food Processing*, **42** (13) (1981) 82–83.

De Michelis, A. and Calvelo, A., Production rate optimization in continuous fluidized bed freezers, *J. Food Sci.*, **50** (1985) 669–673.

De Michelis A. and Calvelo, A., Longitudinal dispersion coefficients for the continuous fluidization of different shaped foods, *J. Food Eng.*, **21** (1994) 331–342.

Ditchev, S. and Richardson, P., Intensification of freezing, in: Oliveira, F.A.R. and Oliveira, J.C., (eds.), Processing foods: quality optimisation and process assessment, CRC Press, Boca Raton, Florida, 1999, 145–162.

Fennema, O.R., Powrie, W.D. and Marth, E.H., Low-temperature preservation of foods and living matter, Marcel Dekker, New York, 1973.

Fikiin, A.G., New method and fluidized water system for intensive chilling and freezing of fish, *Food Control*, **3** (1992) 153–160.

Holdsworth, S.D., Preservation of fruits and vegetable food products, Macmillan, Basingstoke, 1983.

Holdsworth, S.D., Physical and engineering aspects of food freezing, in: Thorne, S., (ed.), Developments in food preservation, volume 4, Elsevier, London, 1987, 153–204.

Jowitt, R., Heat transfer in some food processing applications of fluidisation, *Chem. Engnr.*, **November** (1977) 779–782.

Kay, J.M. and Nedderman, R.M., Fluid mechanics and transfer processes, Cambridge University Press, Cambridge, 1985.

Kelly, M., (1965) cited by: Holdsworth, S.D., Physical and engineering aspects of food freezing, in: Thorne, S., (ed.), Developments in food preservation, volume 4, Elsevier, London, 1987, 153–204.

Kunii, D. and Levenspiel, O., Fluidization engineering, Butterworth-Heinemann, Oxford, 1991.

Marin, M., Rios, G.M. and Gibert, H., Freezing of food products in a fluidized flotation cell, *J. Powder Bulk Solid Tech.*, **7** (1983) 21–25.

McAdams, W.H., Heat transmission, McGraw-Hill, New York, 1954.

Mermelstein, H., Fluidized bed freezer provides efficient IQF processing of seafood, *Food Technology*, **52** (1998) 72.

Morrison, C.R., Fish and shellfish, in: Mallett, C.P., (ed.), Frozen food technology, Blackie, Glasgow, 1993.

Nagaoka, J., Takagi, S. and Hotani, S., Experiments on the freezing of fish in an air-blast freezer, *Proc. 9th Int. Congr. Refrig.*, **2** (1955) 4.

Persson, P.O., Fluidizing technique in food freezing, *ASHRAE J*, **June** (1967) 42–44.

Persson, P.O. and Londahl, G., Freezing technology, in: Mallett, C.P., (ed.), Frozen food technology, Blackie, Glasgow, 1993.

Pham, Q.T., Simplified equation for predicting the freezing time of foodstuffs, *J. Food Tech.*, **21** (1986) 209–219.

Plank, R., Die Gefrierdauer von Eisblocken [The freezing period of ice blocks], *Zeitschrift fur die gesamte Kalte-Ind.*, **20** (6) (1913) 109–114.

Plank, R., Beitrage zur Berechnung und Bewertung der Gefriergesch Windigkeit von Lebensmitteln [Contribution to the estimation and assessment of the freezing time of foods], *Beiheft zur Zeitschrift fur die gesamte Kalte-Ind.*, Reihe 3, Heft 10 (1941) 1–16.

Poulsen, K.P., Energy in the food freezing industry, in: Singh, R.P., (ed.), Energy in food processing, Elsevier, Amsterdam, 1986, 155–177.

Rabinovich, L.B., Sechenov, G.P. and Al'tshuler, V.N., (1967), cited by: Botterill, J.S.M., Fluid-bed heat transfer, Academic Press, London, 1975.

Rao, M.A. and Rizvi, S.S.H., Engineering properties of foods, Dekker, New York, 1995.

Reynoso, R.O. and Calvelo, A., Comparison between fixed and fluidised bed continous pea freezers, *Int. J. Refrig.*, **8** (1985) 109–115.

Rios, G.M., Marin, M. and Gibert, H., New developments of fluidisation in the IQF food area, in: McKenna, B.M., (ed.), Engineering and food, volume 2: processing applications. Third International Congress on Engineering and Food, Dublin 1983, Elsevier, London, 1984, 669–677.

Sheen, S. and Whitney, L.F., Modelling heat transfer in fluidized beds of large particles and its applications in the freezing of large food items, *J. Food Eng.*, **12** (1990) 249–265.

Singh, R.P. and Heldman, D.R., Introduction to food engineering, Academic Press, London, 2001.

Smith, P.G., Introduction to food process engineering, Kluwer Academic, New York, 2003.

Tressler, D.K., van Arsdel, W.B. and Copley, M.J., (eds.), The freezing preservation of foods, volume 1, principles of refrigeration equipment for food freezing refrigerating and transporting frozen foods, 4th ed., AVI, Westport, Connecticut, 1968.

Vazquez, A. and Calvelo, A., Gas particle heat transfer coefficient in fluidized pea beds, *J. Food Proc. Eng.*, **4** (1980) 53–70.

Vazquez, A. and Calvelo, A., Gas particle heat transfer coefficient for the fluidization of different shaped foods, *J. Food Sci.*, **48** (1983a) 114–118.

Vazquez, A. and Calvelo, A., Modelling of residence times in continuous fluidized bed freezers, *J. Food Sci.*, **48** (1983b) 1081–1085.

Chapter 4
Drying

Introduction

Despite the vastly increased importance and popularity of frozen food in recent years, the drying of food remains a very widespread operation in the food industry both for the preservation of foods in their final form and as an intermediate operation; according to Mujumdar (1997), drying accounts for 10% of all energy consumption in the food industry. The process of removing moisture from foods is referred to as both drying and dehydration. Barbosa-Cánovas and Vega-Mercado (1996) distinguish between 'dehydrated food', referring to foods with less than 2.5% moisture on a dry weight basis, and 'dried food' containing more than 2.5% moisture. The term 'drying' will be used throughout this chapter to mean the removal of water from foods using thermal energy.

Drying may be viewed as either a preservation technique or as a manufacturing step and in many cases performs both functions simultaneously. As a food preservation method, drying implies a reduction in the moisture content of a food to a level where the growth of microorganisms is inhibited or where the rate of an adverse chemical reaction is minimised. Beyond its use as a preservation process, drying is an operation which may significantly change the physical form of a product, for example the formation of flakes or granules or the production of a powdered solid from a liquid, as in the spray drying of milk. Drying is used to improve the material handling properties and the flow of food powders and to reduce overall product weight and therefore generates a reduction in transport costs. A major disadvantage of drying is the operating cost associated with the required thermal energy input.

In a gas-solid fluidized bed drier the heat for evaporation of the solvent is supplied in the inlet gas, usually air. The rapid particle mixing and high heat fluxes which characterise aggregative fluidization are exploited to give high drying rates, close temperature control (which avoids overheating and thermal damage) and a uniform moisture content. Despite these significant advantages, it can be difficult to operate a fluidized bed drier successfully. First, it is essential that wet food particles entering the bed are capable of being fluidized immediately. If the inlet moisture content is too high then significant agglomeration may occur and the particles may reach a size at which their minimum fluidizing velocity exceeds the superficial velocity in the bed. In addition, clumps of wet material are likely to adhere to the internal surfaces of the bed and block the distributor plate, factors which lead to a loss of fluidization and the inevitable shut-down of the drier. Many food particles are inherently sticky and cohesive and thus difficult to fluidize (Shilton and Niranjan, 1993). Even without the problem of excessive inlet moisture content, a feed with a wide particle size distribution may not dry uniformly, and it is difficult to maintain constant hydrodynamic conditions (Romankov, 1971). Shilton and Niranjan (1993) suggested that fast fluidization has the potential for drying fine particles because of the very high mass transfer rates obtainable. Second, sufficient heat must be supplied to the bed to evaporate the moisture in the feed and satisfy the material and energy balances. If this is not the case, and the exhaust gas from the bed becomes saturated, a phenomenon known as quenching (Nienow and Rowe, 1975; Smith and Nienow, 1983) occurs which in turn results in defluidization of the bed. Quenching is discussed at greater length in Chapter 5. In addition to these basic considerations, problems may occur with the generation of static electricity (Romankov, 1971) and drying offers considerable potential for dust explosions, particularly in the freeboard of the bed (Bahu, 1997).

Despite these limitations, fluidized bed drying is used extensively for a very wide range of food applications including fruits, diced vegetables, grain and sugar; these are listed in greater detail in Table 4.1.

Table 4.1 Some reported applications of fluidized bed drying of foods.

Peas	Hatamipour and Mowla, 2003; Harjinder and Bawa, 2002
Carrots	Planinic *et al.*, 2005; Prakash *et al.*, 2004; Reyes *et al.*, 2002
Green beans	Senadeera *et al.*, 2000
Peppers	Kaymak-Ertekin, 2002
Sliced celery	Efremov and Kudra, 2004
Chopped alfalfa	Keey, 1997
Brewer's yeast	Luna-Solano *et al.*, 2003
Paddy rice	Madhiyanon and Soponronnarit, 2005; Rordprapat *et al.*, 2005
Maize	Prachayawarakorn *et al.*, 2004
Chopped coconut	Niamnuy and Devahastin, 2005
White sugar	Pakowski and Grochowski, 1997
Sugar beet pulp	Krell *et al.*, 2003
Cranberries	Grabowski *et al.*, 2002
Food colours	White, 1983
Tea	Temple *et al.*, 2001

There are a few reported cases of drying food pieces in a fluidized bed of inert particles such as strawberries in a bed of sugar and potatoes in a bed of salt (Shilton and Niranjan, 1993). Hatamipour and Mowla (2002) dried cylindrical carrot pieces in a bed of glass spheres. This approach to overcoming the difficulty of fluidizing very large food pieces can lead to problems both with segregation and sticking of the inert particles to the food (Shilton and Niranjan, 1993).

Principles of drying

Water activity

Before proceeding further it is important to note that the moisture content of food is often expressed in terms of water activity a_W which is defined as the partial pressure of water vapour above the food surface p_w divided by the pure component vapour pressure of water p'_w at the same temperature as the sample, and therefore

$$a_W = \frac{p_w}{p'_w} \qquad 4.1$$

The partial pressure p_w can be obtained from Raoult's law which relates the partial pressure of water vapour above the surface of an aqueous solution held at a constant temperature, the mole fraction of water in the solution x_w and the pure component vapour pressure of water, i.e.

$$p_w = x_w p'_w \qquad 4.2$$

where x_w is defined in relation only to the soluble constituents within a food and ignores the insoluble components. Equation 4.1 must be true for the extreme values of the concentration of water, i.e. for $x_w = 1$, when the liquid phase is pure water and the partial pressure is equal to the vapour pressure, and for $x_w = 0$, when the water content is zero and there is no partial pressure of water in the vapour phase. Raoult's law generally holds only at relatively high values of x_w and beyond this range applies only to ideal systems. Water is not an ideal material and it is necessary to introduce an activity coefficient γ such that

$$p_w = \gamma x_w p'_w \qquad 4.3$$

Substituting into equation 4.1 gives

$$a_W = \frac{\gamma x_w p'_w}{p'_w} \qquad 4.4$$

and thus

$$a_W = \gamma x_w \qquad \text{4.5}$$

For an ideal system this reduces to

$$a_W = x_w \qquad \text{4.6}$$

Thus for both an ideal system and for high moisture contents, water activity is effectively equal to the mole fraction of water in the liquid phase within the food. There are difficulties in using equation 4.5 because activity coefficients are complex functions of both temperature and moisture content and are not readily determined. Now, relative humidity is defined by

$$\%RH = \frac{p_w}{p'_w} \times 100 \qquad \text{4.7}$$

If the food is in contact with, and at thermal equilibrium with, the surrounding air then, from equation 4.1, water activity is equal to the fractional relative humidity. That is

$$a_W = \frac{\%RH}{100} \qquad \text{4.8}$$

This now forms the basis of the experimental measurement of water activity.

Effect of water activity on microbial growth

Water activity affects both the logarithmic and the stationary phases of microbial growth (Barbosa-Cánovas and Vega-Mercado, 1996) and there is a range of water activity for the growth of all micro-organisms which is widest at the optimum growth temperature for a given species. There are significant differences in the effect of water activity on bacteria, yeasts and moulds: bacteria generally require a greater water activity or moisture content for growth than do fungi; yeasts and moulds tend to grow over a wider range of water activity than do bacteria. The lower limit to water activity for spoilage bacteria is approximately 0.90 although for some bacteria (e.g. *Clostridium botulinum* and pseudomonads) this may be as high as 0.98 (Barbosa-Cánovas and Vega-Mercado, 1996) and as low as 0.75 for halophilic bacteria, i.e. those which can tolerate a high salt content. For yeasts the lower limit is about 0.85–0.88 but is as low as 0.60 for osmophilic yeasts which thrive with high osmotic pressures. The minimum water activity for moulds is generally lower than that of either bacteria or yeasts at approximately 0.80 but is lower still for xerophilic moulds at about 0.60. Forsythe (2000) gives more detailed information for a wide range of micro-organisms.

All of these figures should be viewed in the context of the water activity of fresh food which is very often greater than 0.99. The rate of chemical reactions in food decreases much more slowly with reduced moisture content and enzymatic activity in foods may be significant at water activities as low as 0.30.

Effect of drying on food structure

After the removal of water from a solid food in order to prolong its shelf-life, the original structure, appearance and taste should be reproduced as far as possible when water is added and the dried food is reconstituted. However, the disadvantages of drying food include a number of very significant physical, chemical and biochemical changes; these are reviewed by Barbosa-Cánovas and Vega-Mercado (1996). The removal of moisture will affect the physical appearance such as colour and texture and also more complex matters such as the retention of nutrients and flavours. The structure of fruit and vegetable tissue may be damaged during drying, resulting in a change to the appearance of foods and difficulties in regaining the original properties on rehydration. The temperature during drying and the rate of water removal have a significant effect on the quality of the dried product but equally, the structure of the raw food may itself influence the mechanism by which water is removed and therefore influence the rate of drying.

Isotherms and equilibrium

There are limits to the amount of water which can be removed from a food material which depend upon the nature of the food and the condition of the ambient air responsible for drying. If the food is dried using a flow of warm dry air then, for a given relative humidity, it is not possible to dry food below its equilibrium moisture content. The equilibrium moisture content is the moisture content of a material when in equilibrium with the partial pressure of water vapour in the surroundings; this quantity varies widely for different materials. Figure 4.1 shows the form of a typical sorption isotherm which is generated by measuring the equilibrium moisture content as a function of water activity. Sorption isotherms are characterised by a sigmoidal curve and

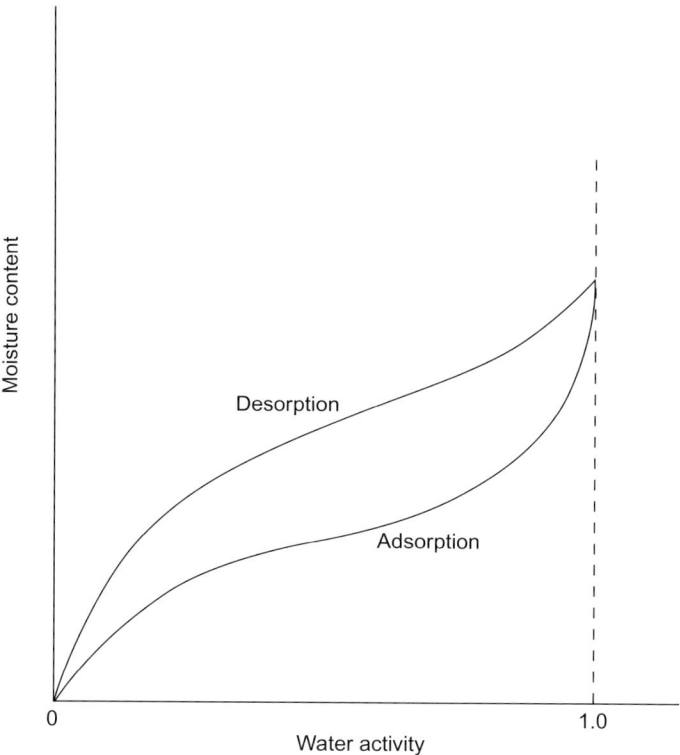

Figure 4.1 Sorption isotherm. From Smith, P.G., Introduction to food process engineering, Kluwer, 2003, figure 12.9. With kind permission of Springer Science and Business Media.

by hysteresis which is produced because of differences in the adsorption and desorption of water. If now the percentage relative humidity of the surrounding air is plotted against the food moisture content then Figure 4.2 results. This shows that as the relative humidity of the air falls there is a corresponding decrease in the equilibrium moisture content of the food and therefore a decrease in the minimum possible final moisture content. Figure 4.2 illustrates the definitions of three other important terms. First, free moisture is any water in excess of equilibrium moisture; it is only free moisture that may be evaporated. Second, bound moisture is water which exerts a vapour pressure less than that of free water of the same temperature; in other words, it has a water activity less than unity, such as water held in small capillaries. Barbosa-Cánovas and Vega-Mercado (1996) subdivide bound moisture into tightly bound ($a_W < 0.3$), moderately bound (a_W 0.3–0.7) and loosely bound ($a_W > 0.7$). Unbound moisture is defined as any water in excess of that which is bound.

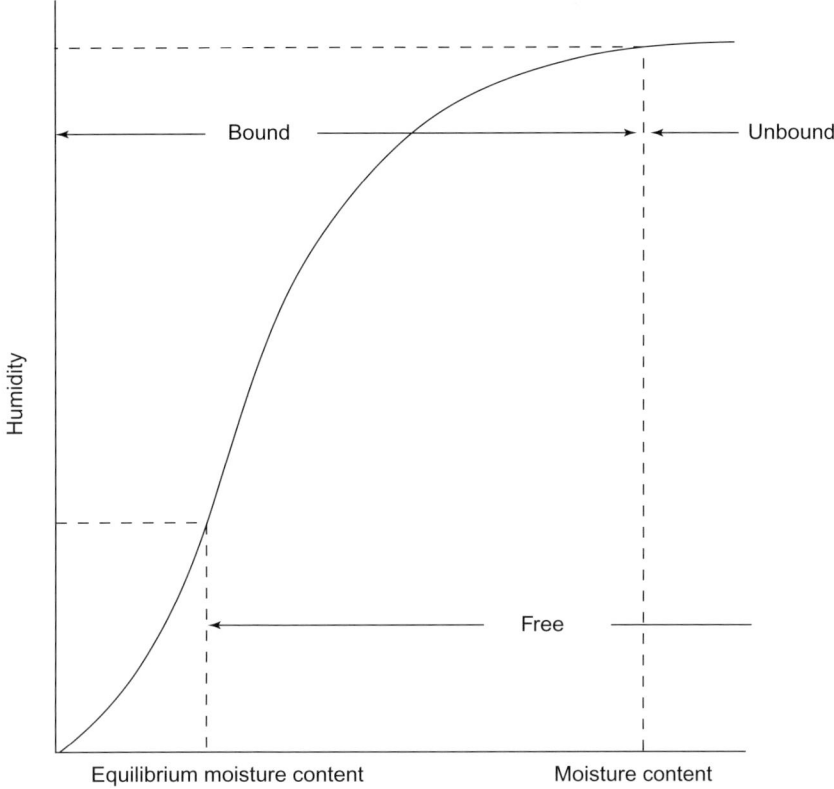

Figure 4.2 Relationship between moisture content and humidity. From Smith, P.G., Introduction to food process engineering, Kluwer, 2003, figure 12.10. With kind permission of Springer Science and Business Media.

Drying kinetics

In a batch-drying process the rate of drying varies with the moisture content of the material and is usually considered to consist of two distinct stages. In the constant rate period (Figure 4.3) the rate of drying is constant from the initial moisture content X_1 down to the critical moisture content X_c. Drying takes place from a saturated surface; vaporised water molecules diffuse through a thin stagnant film of gas close to the surface of the material, which contains all the resistance to mass transfer, before being transported into the bulk of the air stream. The drying rate therefore is not a function of the material being dried but rather a function of the mass transfer characteristics in the surrounding gas stream because the water is vaporised from what is effectively a free water surface.

The rate of drying R is given by

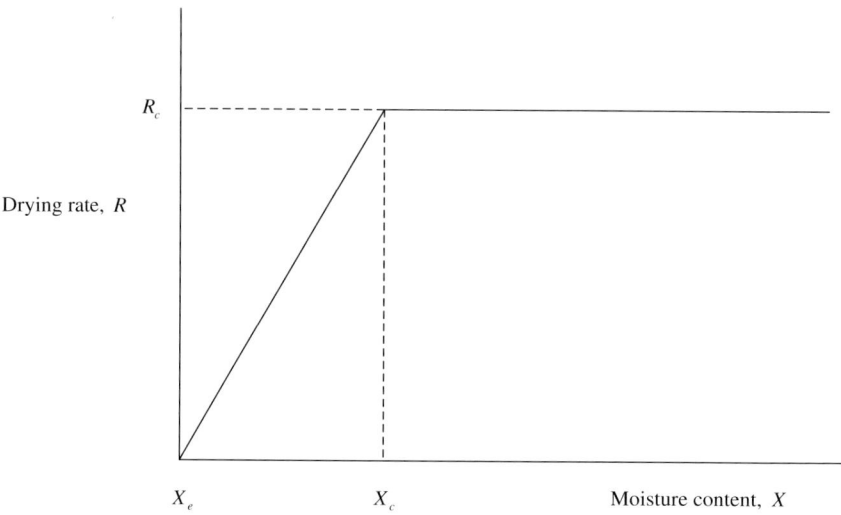

Figure 4.3 Batch drying: drying rate versus moisture content.

$$R = -\frac{W}{A}\frac{dX}{dt} \qquad\qquad 4.9$$

where W is the mass of dry solid, A is the area over which drying takes place and $\dfrac{dX}{dt}$ represents the rate of change of the moisture content on a dry basis. In the constant rate period, where both the initial and final (X_2) moisture contents are greater than X_c, the batch drying time t then becomes

$$t = \frac{W}{AR_c}(X_1 - X_2) \qquad\qquad 4.10$$

where R_c is the constant rate of drying.

The falling rate period is that which is observed at moisture contents below X_c. The rate of drying decreases because the food surface is no longer capable of supplying sufficient free moisture to saturate the air above it; the particle surface becomes dry and the water interface moves further into the particle. The rate of drying is then influenced by the mechanism of transport of moisture from within the particle to the surface and becomes increasingly slow and independent of conditions at the surface. Both capillary forces and the diffusion of water vapour through the porous structure of the food particle may control the movement of water to the particle surface and thus govern the drying rate (Reay and Baker, 1985). Certainly mass transfer through the solid phase is very slow, particularly at low moisture contents, and difficult to predict because of the uncertainty of determining the effective diffusivity of water vapour within the solid phase.

For the falling rate period the batch drying time is given by

$$t = \frac{W}{A} \int_{X_2}^{X_1} \frac{dX}{R}$$ 4.11

which must be solved by determining the relationship between drying rate and moisture content from experiment. However, in some cases it may be reasonable to assume a linear relationship between R and C and thus

$$t = \frac{W}{A} \int_{X_2}^{X_1} \frac{dX}{mX + b}$$ 4.12

where m and b are constants, which results (Smith, 2003) in

$$t = \frac{W}{AR_c}(X_c - X_e)\ln\left(\frac{X_1 - X_e}{X_2 - X_e}\right)$$ 4.13

where X_e is the equilibrium moisture content.

Reay and Baker (1985) point out that because few particles will fluidize with significant surface moisture present, the constant rate period may be extremely short and with non-porous materials the falling rate period may not be observable.

Classification of driers

The fluidized bed drier is one amongst a very large number of industrial drier types available. The classification of driers for food processing is difficult because of the huge variety and the large number of process variables involved. For example, fluidized bed driers can be operated in both batch and continuous modes with all the usual disadvantages associated with batch operation: higher labour costs and a batch-to-batch variation in product quality. On the other hand, a continuous drier can be integrated into a continuous process and should result in lower unit costs at high tonnages. A further classification is as a direct drier, i.e. one in which wet solids are exposed directly to the hot gases which provide the latent heat of vaporisation, rather than as an indirect drier in which there is a physical barrier between the food being dried and the heat source. A disadvantage of direct driers is the potentially considerable loss of material because small dried particles are carried away in a high-velocity exhaust gas stream. However, White (1983) suggested that direct gas firing of a fluidized bed drier for food materials was very economical, given fuel costs at the time. Elutriation, of course, is a particular disadvantage of fluidization. Finally, a fluidized bed drier can be thought of as a through-circulation drier in which the hot gas passes through the solid, as opposed to a cross-circulation drier in which the drying medium passes over the surface of the solids, as in a tray drier. A fuller discussion of the classification of driers can be found in Romankov (1971) and in Mujumdar (1997).

Fluidized bed drying

Material and energy balances

An overall material balance across a fluidized bed drier may be written as follows:

$$\text{dry air in} + \text{water vapour in} + \text{dry solids in} + \text{moisture in}$$
$$= \text{dry air out} + \text{water vapour out} + \text{dry solids out} \qquad 4.14$$
$$+ \text{moisture out}$$

and referring to the schematic diagram in Figure 4.4 this becomes

$$G + H_i G + W + X_1 W = G + H_o G + W + X_2 W \qquad 4.15$$

where, assuming that the fluidizing gas is air, G is the mass flow rate of dry air entering the bed, H_i and H_o are the absolute humidities of the fluidizing air and exhaust air respectively; and the other symbols are as defined above. Removing the dry air and dry solids flow rates from each side of equation 4.15 results in the component material balance for water

$$W(X_1 - X_2) = G(H_o - H_i) \qquad 4.16$$

Equation 4.16 indicates that the feed rate of wet solids must be below that which would result in saturation of the exhaust air stream. Failure to satisfy the moisture balance leads to wet quenching of the bed (see Chapter 5).

The energy balance is now

$$Gc_A(T_i - 0) + H_i Gc_{vap}(T_i - 0) + Wc_S(T_F - 0) + X_1 Wc_w(T_F - 0) + Q_W$$
$$= Gc_A(T_o - 0) + H_o G[c_{vap}(T_o - 0) + h_{fg}] \qquad 4.17$$
$$+ Wc_S(T_S - 0) + X_2 Wc_w(T_S - 0) + Q_L$$

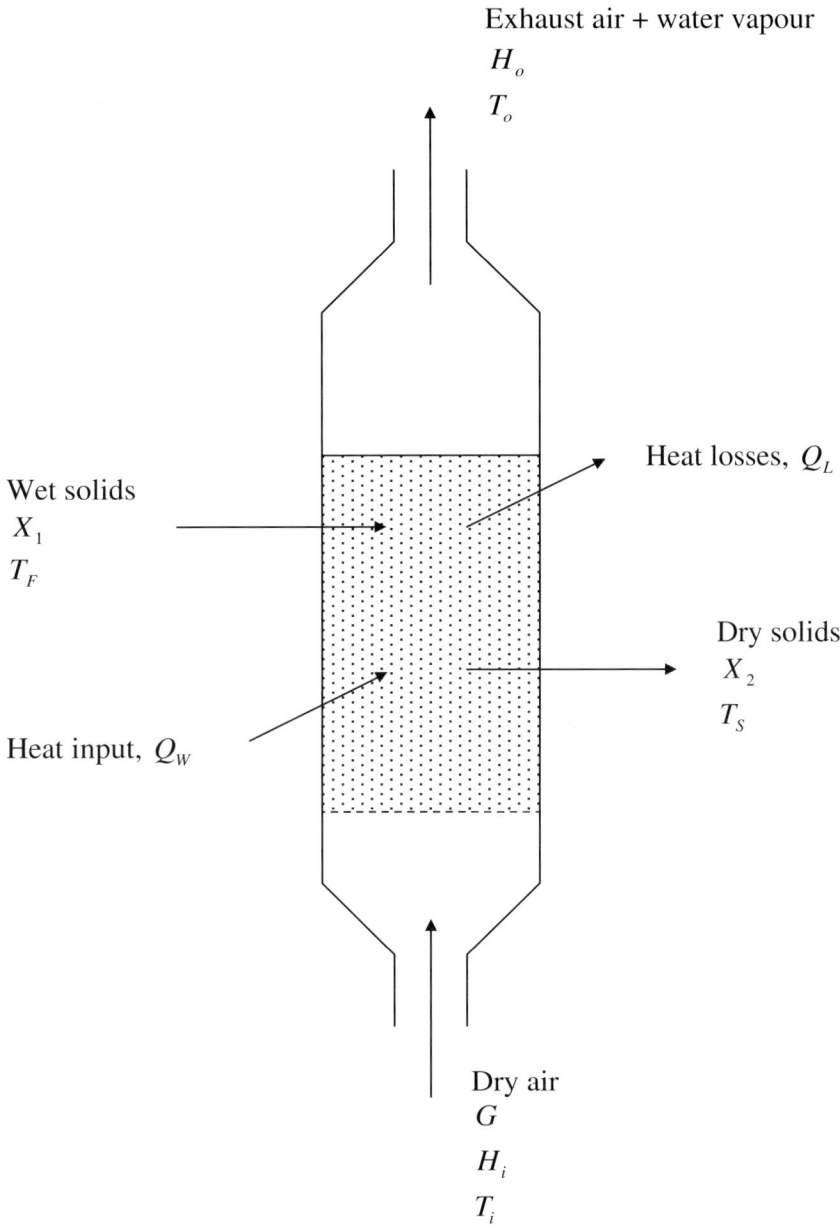

Figure 4.4   Fluidized bed drier: material and enthalpy balance.

where the respective terms represent the energy flows associated with the terms in equation 4.14, with the addition of Q_W, the rate at which heat may be supplied through the bed wall or from an immersed coil where this is present, and Q_L the heat losses from the system.

Both the enthalpy change in the solids and associated moisture and the enthalpy of the water vapour in the fluidizing air are relatively small. Neglecting these terms simplifies the energy balance to

$$Gc_A(T_i - T_o) + Q_W = H_oG[c_{vap}(T_o - 0) + h_{fg}] + Q_L \qquad 4.18$$

which, because the sensible heat of the water vapour in the exhaust gas is small compared to the latent heat of vaporisation, and because the heat for evaporation is usually supplied only by the fluidizing gas, can be reduced further to

$$Gc_A(T_i - T_o) \cong H_oGh_{fg} + Q_L \qquad 4.19$$

Equation 4.19 indicates that the enthalpy change of the fluidizing gas is approximately equal to the sum of the latent heat of the water which is evaporated from the wet solids feed and the heat losses from the bed.

The well-mixed drier

There are two fundamentally different designs of fluidized bed: the well-mixed bed and the plug flow bed. In addition, various combinations of fluidized bed drier may be used to meet particular applications; Kunii and Levenspiel (1991) illustrate a series of such designs (see Variations in fluidized bed drier design, below).

The continuously operated well-mixed bed fluidized bed drier, sometimes referred to as a CSTR, is based on the concept of the continuous stirred tank reactor. It is either circular or square in cross-section and relatively deep; Romankov (1971) and Bahu (1997) suggest typical bed depths of up to 0.4 m. The particle residence time distribution approximates to perfect mixing, i.e. the outlet stream will have the same concentration, or moisture content, as the average moisture content in the bed. The bed temperature is uniform and equal to the exhaust gas temperature; thermal efficiency is high. This type of drier has the great advantage of being able to accept very wet feed particles, although perhaps too wet to be fluidized themselves, because they are surrounded immediately by much drier material and thus fluidization is maintained. However, the overall material balance must still be satisfied.

The well-mixed drier has the distinct disadvantage of not being able to dry particles to very low moisture contents because of the wide particle residence time distribution. In other words, some particles leave the bed immediately, having had no opportunity to dry and effectively bypass the bed; some circulate within the bed for a long period and dry completely; and there is a range of times and moisture contents between these extremes. Thus the particles leaving the bed have a wide range of moisture content and the average moisture content is well above the minimum. The prediction of the average outlet moisture content X_o from a continuous well-mixed drier, and therefore the design of such equipment, is based on the model of Vanacek *et al.* (1966) who proposed

$$X_o = \int_0^\infty E(t)X(t)dt \qquad 4.20$$

where $X(t)$ is the moisture content on dry basis as a function of time t and is obtained from an experimentally determined drying curve. $E(t)$ is the residence time distribution of particles in the bed which is given by

$$E(t) = \frac{1}{t_m}\exp\left(-\frac{t}{t_m}\right) \qquad 4.21$$

where t_m is the mean residence time in the bed which is equal to the bed mass hold-up divided by the mass feed rate, both on a dry weight basis.

Bahu (1997) suggests obtaining a drying curve at a constant bed temperature (which he terms an isothermal bed batch-drying curve or IBBDC). Measuring such a curve experimentally is difficult because, with a constant gas inlet temperature, the bed temperature falls as the batch-drying test proceeds and the drying rate decreases. Thus it becomes necessary to progressively reduce the inlet gas temperature. It is more convenient to measure what Bahu calls an isothermal inlet gas batch-drying curve (IIGBDC). However, Bahu (1997) and Reay and Baker (1985) describe a model, due originally to Reay and Allen (1982), which allows an IBBDC to be constructed from an IIGBDC.

Reay and Allen (1982) assumed that the drying rate is dependent upon the capacity of the gas in the dense phase to absorb moisture and therefore that the drying rate is proportional to the difference between the partial pressure of water vapour p_x in equilibrium with a moisture content X and the partial pressure p_i in the inlet gas which is in equilibrium with the equilibrium moisture content of the particles X_e. Therefore the drying rate at a moisture content X is given by

$$\frac{dX}{dt} \propto (p_x - p_i) \qquad 4.22$$

Now because these authors further assumed a linear desorption isotherm, i.e. a linear relationship between the equilibrium partial pressure of water vapour and the solids moisture content, it follows that

$$\frac{dX}{dt} \propto (p_S - p_i)\frac{(X - X_e)}{(X_S - X_e)} \qquad 4.23$$

where p_S is the saturated vapour pressure of water and X_S is the moisture content with which p_S is in equilibrium. Further, if X_S is very much larger than X_e, and thus $X_S - X_e$ is nearly constant over the range of temperature encountered, equation 4.23 reduces to

$$\frac{dX}{dt} \propto (p_S - p_i)(X - X_e) \qquad 4.24$$

Applying this relationship to drying curves obtained at two temperatures T_1 and T_2, for a given change in moisture content ΔX, gives

$$\frac{(\Delta t)_{T_2}}{(\Delta t)_{T_1}} = \frac{[(p_S - p_i)(X - X_e)]_{T_1}}{[(p_S - p_i)(X - X_e)]_{T_2}} \qquad 4.25$$

Reay and Allen (1982) and Reay and Baker (1985) used equation 4.25 to generate drying curves, in increments of Δt, at a temperature T_2 from an experimental drying curve obtained at a temperature T_1. They present data from the drying of a range of materials, including wheat, which show that predicted and experimental drying curves are in very good agreement and that the method can be used to generate an isothermal bed batch-drying curve from an isothermal inlet gas batch-drying curve. Using equation 4.25, time intervals Δt obtained for increments in moisture content ΔX at T_1 (IIGBDC) are used to calculate the time intervals at T_2 (IBBDC) needed for the same increment ΔX. Bahu (1997) reports an extension of this method to take into account the effect of inlet gas humidity.

Reay and Baker (1985) suggest that there are two types of particle which are encountered in fluidized bed drying: type A particles which dry easily and where the water vapour in the gas leaving the bed is in equilibrium with the moisture in the particle; and type B particles which have a high internal resistance to the mass transfer of water and therefore dry very slowly. Wheat is an example of the latter (Giner and Calvelo, 1987).

For type A solids, drying takes place close to the distributor, perhaps within 2–5 cm of the plate. Increasing the bed depth beyond this level has no effect on the humidity of the exhaust gas which is in any case well below the humidity in equilibrium with the moisture in the bed particles. The total gas flow is well below the equilibrium humidity and is made up of gas leaving the dense phase which is close to equilibrium with the particles and gas leaving the bubble phase which is not. The drying rate increases in proportion to gas velocity and thus all the gas is assumed to be in good contact with the particles in the region immediately above the distributor plate. This is confirmed by the experimental observation that particle size has very little effect on the drying rate.

With type B solids, Reay and Baker suggest that bed depth does have a significant effect: increasing bed depth increases the humidity of the outlet gas. Therefore, in these cases, even the dense-phase gas is not at equilibrium with the moisture in the bed particles. In contrast to type A solids, gas velocity has no effect on the drying rate but there is a significant effect of particle size: drying rate is proportional to the diameter squared. These observations are explained by the high internal resistance to the movement of water and the slow diffusion of water to the particle surface.

The plug flow drier

An alternative to the CSTR arrangement is a bed based on the principle of the plug flow (PF) reactor. In its simplest form, a PF fluidized bed drier consists of a long, narrow rectangular cross-section container (aspect ratios between 5 and 40 are common) with the feed inlet at one end and a weir at the other over which dried solids flow out to the next process stage (Figure 4.5). The depth of the bed is usually less than about 10–15 cm, to reduce the possibility of back-mixing, and is

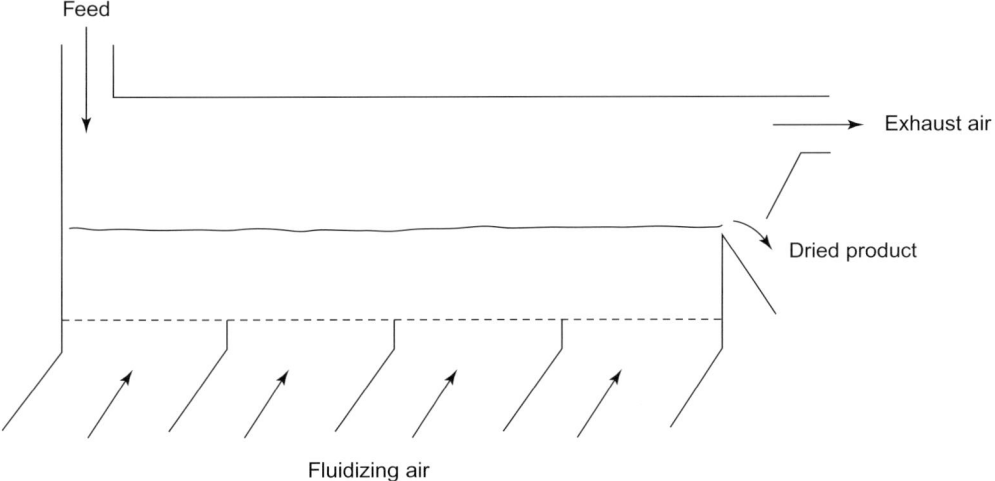

Figure 4.5 Schematic diagram of plug flow fluidized bed drier.

controlled by the weir height. Therefore, for a constant feed rate, the average residence time in the drier can be controlled. It is an assumption of plug flow that very little longitudinal mixing, or back-mixing, takes place and therefore the bed is characterised by a narrow residence time distribution and a more uniform outlet moisture content; the uniformity of product moisture content increases with the length/width ratio (Bahu, 1997). A particular advantage of this kind of drier is the ability to obtain a very low final moisture content. However, this is offset by the requirement for a lower inlet moisture content than with a well-mixed bed and the need for the feed to be instantly fluidizable. Bahu (1997) suggests that increasing the bed area per unit bed mass helps to break up the wet feed and disperse it away from the inlet. In addition, a PF bed is much less thermally efficient than a well-mixed bed. If the inlet gas temperature is kept constant then the bed temperature rises along the bed length and therefore the gas at the discharge end is responsible for relatively little drying.

Some of these disadvantages can be alleviated by changing the processing conditions as the particles move through the bed. The provision of separate compartments in the plenum chamber allows different gas velocities and temperatures to be used along the bed length. For example, at the bed inlet higher velocities and temperatures can be used to cope with the higher moisture content of the feed whilst more moderate conditions can be employed as the moisture content is reduced. Lower gas temperatures at the outlet of the bed will increase the thermal efficiency and allow the particles to be cooled prior to discharge.

For ideal plug flow, the residence time in the bed corresponds with the batch-drying time obtained from an IIGBDC for the same food material and with the same fluidizing gas velocity and bed height. It is therefore relatively straightforward to design a fluidized bed drier, based on a mean residence time, in such cases. However, where there is deviation from ideal PF behaviour, i.e. when there is a degree of longitudinal mixing, the distribution of residence times is given by equation 4.26

$$E(t) = \frac{1}{2\sqrt{\pi B}} \exp\left[\frac{-\left(1 - \frac{t}{t_m}\right)^2}{4B} \right] \qquad 4.26$$

where B is the axial dispersion number which is defined by

$$B = \frac{D_z t}{L^2} \qquad 4.27$$

and D_z is the effective diffusivity of particles in the direction of solids flow and L is the bed length. Longitudinal mixing in the bed can be described by Fick's second law of diffusion for unsteady-state mass transfer in one dimension (in the same way as it was used to describe vertical dispersion in the bed in Chapter 2). Thus

$$\frac{\partial C_S}{\partial t} = D_z \frac{\partial^2 C_S}{\partial z^2}$$ 4.28

in which C_S is the concentration of a group of particular marked or tagged particles at a given distance along the length of the bed z. Again, this equation assumes that the variation in concentration in the other two orthogonal dimensions is equal to zero.

The axial dispersion number B is therefore a measure of the extent of backmixing. For ideal plug flow $B = 0$, whereas for a perfectly well-mixed system $B = \infty$. Equation 4.26 is valid for small deviations from ideality which is usually assumed to be $B < 0.1$. Vazquez and Calvelo (1983) and De Michelis and Calvelo (1994) measured longitudinal dispersion coefficients in the context of fluidized bed freezing (Chapter 3) and found that D_z was strongly dependent on bed height and gas velocity.

Variations in fluidized bed drier design

There are a number of possible variations in the design of the basic types of fluidized bed drier and various combinations of drier which may be used to meet particular applications. In a plug flow drier the variation in drying conditions along the length of the drier may be achieved either by using separate compartments in the plenum or by extending the partition plates into the bed with solids flowing over each weir into the next compartment. Partitions may also be placed longitudinally in a plug flow bed so as to form a hairpin arrangement (Figure 4.6) which has the effect of broadening the width of the bed but reducing the overall length.

Kunii and Levenspiel (1991) illustrate a series of designs including: (1) a two-stage drier suitable for heat-sensitive materials which require low inlet gas temperatures but incorporating heat recovery from the outlet solids in order to improve thermal efficiency; (2) the supply of heat from heating coils immersed in the bed which has a particular advantage if the gas velocity has to be reduced in order to retain fine particles in the bed; (3) operating the bed at high pressure and fluidizing with superheated steam to give very high thermal efficiencies and allowing the introduction of very wet feed solids. A further advantage of this technique is the generation of low-pressure steam for use

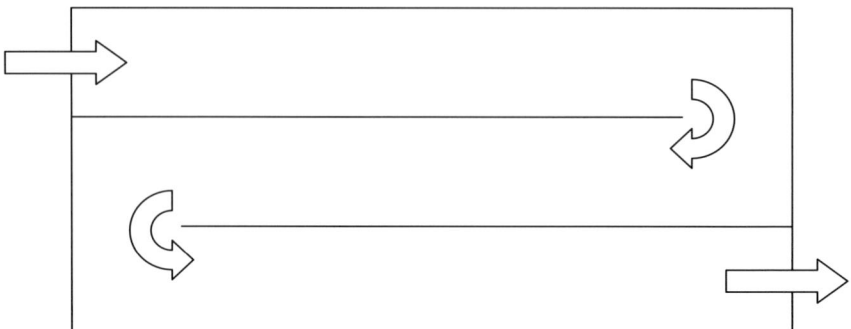

Figure 4.6   Plug flow fluidized bed drier: 'hairpin' arrangement.

in subsequent drying stages. Finally (4), the recirculation of fluidizing steam in a closed circuit is possible and is particularly useful if the solvent is other than water and must be contained.

A number of driers may be used in series. For example, a cascade of well-mixed driers approximates to the performance of a plug flow drier. However, of more practical significance is the combination of a well-mixed drier followed by a plug flow drier for drying very wet solids which may defluidize if fed directly to a PF unit. The CSTR drier is better able to cope with the high moisture content feed and the plug flow drier allows a low final moisture content to be obtained.

Other fluidized bed drying techniques

Vibro-fluidization

A number of techniques have been used to assist the flow of fluidized solids through driers and to prevent excessive agglomeration of wet material. Gravity flow is normally used but it is possible to mount the entire fluidized bed on a spring chassis and vibrate it. Vibrations imposed on a fluidized bed via a vibrating distributor plate can improve the quality of fluidization (Botterill, 1975) and allow operation at velocities below minimum fluidizing gas velocity as well as assisting the flow of solids. Han *et al.* (1991), using wheat, showed that the flow of particles in a continuous vibro-fluidized bed could be considered as plug flow and that vibration intensity was the most significant factor to affect particle mean residence time, particle diffusivity and constant drying rate. In experiments using ground eggshell and a tracer technique, Brod *et al.* (2004) found that the residence time in a vibro-fluidized drier increased as the vibration amplitude decreased and as the flow rate of solids increased, at low solids flow rates and low gas velocities. However, at higher gas velocities the residence time decreased with increasing gas velocity. Frequencies up to 50 Hz and amplitudes of 5 mm have been used. However, there is a limit to the bed length of about 10 m because of flexing of the distributor.

Bahu (1997) suggests that vibro-fluidized beds are useful for wide particle size distributions, because the gas velocity can be maintained below that at which excessive elutriation of the fine particles occurs while the large particles are kept in motion by the vibration. Equally, for narrow size distributions of large particles (1 mm or greater) the gas velocity need only be just above the minimum fluidizing velocity. Soponronnarit *et al.* (2001) developed a vibrating fluidized bed drier for paddy rice (with a capacity up to $5.0 \, t \, h^{-1}$) using a frequency of 7.3 Hz and an amplitude of 5 mm and found that the combined electrical power of the blower and vibrator motors was only 55% of that used by the blower motor alone in a conventional fluidized bed drier. A vibrating fluidized bed drier was found to have the highest energy efficiency for the finish drying of osmotically pre-treated cranberries (Grabowski *et al.*, 2002).

Vibro-fluidization is used for cohesive, sticky solids or friable foods (Bahu, 1997) and for materials which would defluidize in a conventional plug flow drier (Reay and Baker, 1985). Vibration of the bed increases the drying rate due to an increase in the surface-to-bed heat transfer coefficient (Reay and Baker, 1985), particularly below minimum fluidizing velocity. A detailed treatment of the mechanisms of vibration fluidization is given by Reay and Baker (1985).

A wide range of applications of vibro-fluidized bed drying has been reported. These include the drying and instantisation of milk, coffee and other beverages (Reay and Baker, 1985) and the rehumidification of dried dairy products (Schuck *et al.*, 2004). Pan *et al.* (2001) used a vibrating fluidized bed drier containing inert particles to dry soymilk. Teflon particles were found to be superior to glass beads as they gave a higher heat transfer coefficient and allowed higher liquid feed rates. Other foods treated include walnuts and peanuts (Reay and Baker, 1985), cooked rice (Ramesh and Srinivasa-Rao, 1996), sugar (Hoks and Elfrink, 1993), tea (Shah and Goyel, 1995), squash slices (Pan *et al.*, 1999a), cabbage seeds (Kai *et al.*, 1999) and carrot (Le Maguer and Mazza, 1984; Pan *et al.*, 1999b).

Mechanical agitation

Alternative methods of assisting the fluidization of difficult materials include the use of impellers in the bed to break up agglomerated particles, particularly to help initial fluidization in plug flow driers (Reay and Baker, 1985), and the use of a pulsed gas flow (Butcher, 1988). Bahu

(1997) describes the use of a tangential gas inlet through a distributor in the bed wall combined with a rotating agitator mounted at the base of the bed. Because additional drying takes place in the freeboard and in the exhaust ducting above the bed, in a similar way to pneumatic or flash drying, the technique is referred to as spin flash drying.

Centrifugal fluidization

The principles of the centrifugal fluidized bed were outlined in Chapter 1 and Cohen and Yang (1995) have reviewed its use for foods. Mass transfer rates, and therefore drying rates, are higher than in a conventional fluidized bed (Chen *et al.*, 1999) and therefore a smaller plant volume is required for a given throughput. Shi *et al.* (2000) showed that the drying rate increased with increasing superficial gas velocity and particle diameter and decreased with bed rotation speed and the initial bed thickness. Empirical correlations for the gas-solid mass transfer coefficient during drying (Chen *et al.*, 1999) and the heat transfer coefficient (Shi *et al.*, 2000) have been proposed as a function of bed thickness, particle diameter, rotation speed and gas velocity. The main applications of centrifugal fluidized bed drying are pre-drying and the removal of surface moisture (Bahu, 1997), specific examples being the drying of diced vegetables (Hanni *et al.*, 1976), rice (Roberts *et al.*, 1980; Chen *et al.*, 1999), breadcrumbs (Fletcher, 1982), soybeans, green beans and red beans (Chen *et al.*, 1999) and sliced potatoes (Shi *et al.*, 2000).

Spouted bed drying

Spouted bed drying has been used successfully for large particles where the gas flow rates for fluidization would be excessive. Used originally for the drying of wheat, this method has found application with large-diameter particulates which are difficult to fluidize, e.g. peas and beans, and for sticky materials. The advantages are a lower pressure drop across the bed and better gas-solid contact (Franca *et al.*, 1998). Grabowski *et al.* (1997) found that high moisture content bakers' yeast particles behaved as Geldart group C or D particles and that consequently fluidization was very difficult. Pressed yeast samples (with a moisture content of 70% on a wet weight basis) were extruded to give cylindrical particles between 0.8 and 1 mm in diameter and 1.5–3 mm in length. These were then dried in a spouted bed. Particles below 35% moisture were able to be fluidized and these authors therefore suggest a two-stage drying process with a spouted bed followed by a conventional fluidized bed drier. Lima and Rocha (1998) demonstrated the potential of a spouted bed drier for the drying of carioca beans 7.5 mm in diameter and with a sphericity of 0.44.

Microwave drying

A major problem with hot air drying, including fluidized bed drying, is energy inefficiency and the generally slow rate of drying. A relatively recent technique which has been developed to alleviate these disadvantages is the combination of microwave heating and fluidized bed drying (Mermelstein, 1998). With microwave drying it is important to obtain a uniform distribution of microwave energy to ensure even heating. In a conventional domestic microwave oven this is achieved by using a rotating metal stirrer within a metal-lined cavity together with rotation of the food on a turntable. Mermelstein (1998) suggests that the agitation of solids in a fluidized or spouted bed overcomes this problem, thus creating a uniform exposure of food particulates to microwave radiation. Feng and Tang (1998) used a combination of 2450 MHz microwaves and a spouted bed for the finish drying of diced evaporated apples from 24% to 5% moisture. The spouted bed, with an inlet gas velocity of $1.7\,m\,s^{-1}$ and an inlet air temperature of 70°C, was placed in a microwave cavity measuring $393 \times 279 \times 167$ mm. These authors report that the drying time was reduced by more than 80% compared with spouted bed drying alone and that the dried product showed less discolouration and better rehydration properties. Mermelstein (1998) also reports the use of the combination of spouted bed and microwave drying for reducing the moisture content of blueberries from 85% to 10%; drying time was reduced to one-sixth of that for spouted bed drying alone and one-twentieth of the time required in a tray drier. Again this method is reported to reduce colour loss on drying.

Kaensup *et al.* (1998) used a 0.1 m diameter conventional fluidized bed, placed inside a 500 W output microwave oven cavity, to dry 4.5 mm diameter white pepper seed kernels. High gas velocities of 5 or $8\,m\,s^{-1}$ were employed. A slight decrease in drying time was demonstrated compared to fluidized bed drying alone at the same air inlet temperatures. A similar experimental arrangement was used by Goksu *et al.* (2005) to dry 2.4 mm diameter macaroni beads in a 73 mm diameter fluidized bed. The rate of drying increased at higher microwave power levels and the estimated effective diffusivity of water in the bed increased from $4.125 \times 10^{-11}\,m^2\,s^{-1}$ without microwave assistance to $8.772 \times 10^{-11}\,m^2\,s^{-1}$ for combined fluidized bed and microwave drying. The authors also suggest that this drying technique prevents case hardening of the macaroni beads. In all the reported studies a microwave frequency of 2450 MHz was used. The use of radiofrequency drying in combination with fluidized beds has also been reported (Jumah, 2005).

Nomenclature

a_W	water activity
A	area
b	constant
B	axial dispersion number
c_A	heat capacity of air
c_S	heat capacity of dry solid
c_{vap}	heat capacity of water vapour
c_w	heat capacity of water
C_S	concentration of tagged particles
D_z	longitudinal dispersion coefficient
$E(t)$	residence time distribution of particles
G	mass flow rate of dry air entering the bed
h_{fg}	latent heat of vaporisation of water
H_i	absolute humidity of fluidizing air
H_o	absolute humidity of exhaust air
L	bed length
m	constant
p_i	partial pressure in the inlet gas in equilibrium with X_e
p_S	saturated vapour pressure of water
p_w	partial pressure of water vapour above the food surface
p'_w	pure component vapour pressure of water
p_x	partial pressure of water vapour in equilibrium with a moisture content X
Q_L	rate of heat loss
Q_W	rate of heat addition through bed wall
R	rate of drying
R_c	constant drying rate
$\%RH$	percentage relative humidity
t	time
t_m	mean residence time
T	temperature
T_F	solids feed temperature
T_i	inlet air temperature
T_o	outlet air temperature
T_S	solids outlet temperature
W	mass of dry solid
x_w	mole fraction of water
X	moisture content on dry basis
X_c	critical moisture content
X_e	equilibrium moisture content
X_o	average outlet moisture content

X_S moisture content in equilibrium with saturated vapour pressure of water

X_1 initial moisture content

X_2 final moisture content

z distance in direction of solids flow

Greek symbols

γ activity coefficient

References

Bahu, R.E., Fluidized bed dryers, in: Baker, C.G.J., (ed.), Industrial drying of foods, Blackie, London, 1997.

Barbosa-Cánovas, G.V. and Vega-Mercado, H., Dehydration of foods, Chapman and Hall, London, 1996.

Botterill, J.S.M., Fluid-bed heat transfer, Academic Press, London, 1975.

Brod, F.P.R., Park, K.J. and de Almeida, R.G., Image analysis to obtain the vibration amplitude and the residence time distribution of a vibro-fluidized dryer, *Food Bioprod. Proc.*, **82** (2004) 157–163.

Butcher, C., Fluid bed dryers, *Chem. Engr.*, **July** (1988) 16–18.

Chen, G.F., Wang, W., Yan, H. and Wang, X.Z., Experimental research on mass transfer in a centrifugal fluidized bed dryer, *Drying Tech.*, **17** (1999) 1845–1857.

Cohen, J.S. and Yang, T.C.S., Progress in food dehydration, *Trends Food Sci. Tech.*, **6** (1995) 20–25.

De Michelis, A. and Calvelo, A., Longitudinal dispersion coefficients for the continuous fluidization of different shaped foods, *J. Food Eng.*, **21** (1994) 331–342.

Efremov, G. and Kudra, T., Calculation of the effective diffusion coefficients by applying a quasi-stationary equation for drying kinetics, *Drying Tech.*, **22** (2004) 2273–2279.

Feng, H. and Tang, J., Microwave finish drying of diced apples in a spouted bed, *J. Food Sci.*, **63** (1998) 679–683.

Fletcher, J., Food dehydration, Campden Food Preservation Research Association, Chipping Campden, Gloucestershire, 1982.

Forsythe, S.J., The microbiology of safe food, Blackwell, Oxford, 2000.

Franca, A.S., Passos, M.L., Charbel, A.L.T. and Massarani, G., Modeling and simulation of airflow in spouted bed dryers, *Drying Tech.*, **16** (1998) 1923–1938.

Giner, S.A. and Calvelo, A., Modelling of wheat drying in fluidized beds, *J. Food Sci.*, **52** (1987) 1358–1363.

Goksu, E.I., Sumnu, G. and Esin, A., Effect of microwave on fluidized bed drying of macaroni beads, *J. Food Eng.*, **66** (2005) 463–468.

Grabowski, S., Mujumdar, A.S., Ramaswamy, H.S. and Strumillo, C., Evaluation of fluidized versus spouted bed drying of baker's yeast, *Drying Tech.*, **15** (1997) 625–634.

Grabowski, S., Marcotte, M., Poirier, M. and Kudra, T., Drying characteristics of osmotically pretreated cranberries – energy and quality aspects, *Drying Tech.*, **20** (2002) 1989–2004.

Han, W., Mai, B. and Gu, T., Residence time distribution and drying characteristics of a continuous vibro-fluidized bed, *Drying Tech.*, **9** (1991) 159–181.

Hanni, P.F., Farkas, D.F. and Brown, G.E., Design and operating parameters for a centrifugal fluidized bed drier, *J. Food Sci.*, **41** (1976) 1172–1176.

Harjinder, K. and Bawa, A.S., Studies on fluidized bed drying of peas, *J. Food Sci. Tech.*, **39** (2002) 272–275.

Hatamipour, M.S. and Mowla, D., Shrinkage of carrots during drying in an inert medium fluidized bed, *J. Food Eng.*, **55** (2002) 247–252.

Hatamipour, M.S. and Mowla, D., Correlations for shrinkage, density and diffusivity for drying of maize and green peas in a fluidized bed energy carrier, *J. Food. Eng.*, **59** (2003) 221–227.

Hoks, D. and Elfrink, E., A new fluidized bed white sugar drier/cooler, *Zuckerindustrie*, **118** (1993) 465–468.

Jumah, R., Modelling and simulation of continuous and intermittent radio frequency-assisted fluidized bed drying of grains, *Food Bioprod. Proc.*, **83** (2005) 203–210.

Kaensup, W., Wongwises, S. and Chutima, S., Drying of pepper seeds using a combined microwave/fluidized bed dryer, *Drying Tech.*, **16** (1998) 853–862.

Kai, Z., Xun, L. Zhide, C. and Junhong, Y., The thermo-imaging experimental study for the optimization of seeds drying, *Drying Tech.*, **17** (1999) 1935–1945.

Kaymak-Ertekin, F., Drying and rehydrating kinetics of green and red peppers, *J. Food Sci.*, **67** (2002) 168–175.

Keey, R.B., The industrial drying of foods: an overview, in: Baker, C.G.J., (ed.), Industrial drying of foods, Blackie, London, 1997.

Krell, L., Moerle-Heynisch, T. and Wunsch, O., Steam drying in the Uelzen sugar factory: initial operating experiences, *Zuckerindustrie*, **128** (2003) 366–370.

Kunii, D. and Levenspiel, O., Fluidization engineering, Butterworth-Heinemann, Oxford, 1991.

Le Maguer, M. and Mazza, G., An engineering analysis of the drying of carrots in a vibro fluidizer, in: McKenna, B.M., (ed.), Engineering and food, volume 1: engineering sciences in the food industry. Third International Congress on Engineering and Food, Dublin 1983. Elsevier, London, 1984, 235–243.

Lima, A.C.C. and Rocha, S.C.S., Bean drying in fixed, spouted and spout-fluid beds: a comparison and empirical modeling, *Drying Tech.*, **16** (1998) 1881–1901.

Luna-Solano, G., Salgado-Cervantes, M.A., Ramirez-Lepe, M., Garcia-Alvarado, M.A. and Rodriguez-Jimenes, G.C., Effect of drying type and drying conditions over fermentative ability of brewer's yeast, *J. Food Proc. Eng.*, **26** (2003) 135–147.

Madhiyanon, T. and Soponronnarit, S., High temperature spouted bed paddy drying with varied downcomer air flows and moisture contents: effects on drying kinetics, critical moisture content, and milling quality, *Drying Tech.*, **23** (2005) 473–495.

Mermelstein, N.H., Microwave and radio frequency drying, *Food Technology*, **52** (11) (1998) 84–86.

Mujumdar, A.S., Drying fundamentals, in: Baker, C.G.J., (ed.), Industrial drying of foods, Blackie, London, 1997.

Niamnuy, C. and Devahastin, S., Drying kinetics and quality of coconut dried in a fluidized bed dryer, *J. Food Eng.*, **66** (2005) 267–271.

Nienow, A.W. and Rowe, P.N., Fluid bed granulation, Separation Process Services, Harwell, 1975.

Pakowski, Z. and Grochowski, J., Drying of white sugar in a fluid bed: simulation and design of industrial scale dryers-coolers, *Drying Tech.*, **15** (1997) 1881–1892.

Pan, Y.K., Zhao, L.J. and Hu, W.B., The effect of tempering-intermittent drying on quality and energy of plant materials, *Drying Tech.*, **17** (1999a) 1795–1812.

Pan, Y.K., Zhao, L.J., Dong, Z.X., Mujumdar, A.S. and Kudra, T., Intermittent drying of carrot in a vibrated fluid bed: effect on product quality, *Drying Tech.*, **17** (1999b) 2323–2340.

Pan, Y.K., Li, J.G., Zhao, L.J., Ye, W.H., Mujumdar, A.S. and Kudra, T., Performance characteristics of the vibrated fluid bed of inert particles for drying of liquid feeds, *Drying Tech.*, **19** (2001) 2003–2018.

Planinic, M, Velic, D., Tomas, S., Bilic, M and Bucic, A., Modelling of drying and rehydration of carrots using Peleg's model, *Eur. Food Res. Tech.*, **221** (2005) 446–451.

Prachayawarakorn, S., Soponronnarit, S., Wetchacama, S. and Chinnabun, K., Methodology for enhancing drying rate and improving maize quality in a fluidised-bed dryer, *J. Stored Prod. Res.*, **40** (2004) 379–393.

Prakash, S., Jha, S.K. and Datta, N., Performance evaluation of blanched carrots dried by three different driers, *J. Food Eng.*, **62** (2004) 305–313.

Ramesh, M.N. and Srinivasa-Rao, P.N., Drying studies of cooked rice in a vibrofluidised bed drier, *J. Food Eng.*, **27** (1996) 389–396.

Reay, D. and Allen, R.W.K., Predicting the performance of a continuous well-mixed fluid bed dryer from batch tests, in: Ashworth, J., (ed.), Proceedings of the Third International Drying Symposium, volume 2, Birmingham, 1982, 130–140.

Reay, D. and Baker, C.G.J., Drying, in: Davidson, J.F., Clift, R. and Harrison, D., (eds.), Fluidization, 2nd ed., Academic Press, London, 1985.

Reyes, A., Alvarez, P.I. and Marquardt, F.H., Drying of carrots in a fluidized bed. I: Effects of drying conditions and modelling, *Drying Tech.*, **20** (2002) 1463–1483.

Roberts, R.L., Carlson, R.A. and Farkas, D.F., Preparation of a quick-cooking brown rice product using a centrifugal fluidized bed drier, *J. Food Sci.*, **45** (1980) 1080–1081.

Romankov, P.G., Drying, in: Davidson, J.F. and Harrison, D., (eds.), Fluidization, Academic Press, London, 1971.

Rordprapat, W., Nathakaranakule, A., Tia, W. and Soponronnarit, S., Comparative study of fluidized bed paddy drying using hot air and super-heated steam, *J. Food Eng.*, **71** (2005) 28–36.

Schuck, P., Mejean, S., Dolivet, A., Jeantet, R., Pirus, P. and Belan, F., Rehumidification of dairy powders in dynamic phase, *Sciences des Aliments.*, **24** (2004) 383–398.

Senadeera, W., Bhandari, B., Young, G. and Wijesinghe, B., Physical properties and fluidization behaviour of fresh green bean particulates during fluidized bed drying, *Food Bioprod. Proc.*, **78** (2000) 43–47.

Shah, R.M. and Goyel, S.K., Drying characteristics of tea fluidised on a vibrating bed. II: Handling of dust grades and its impact on quality, *Drying Tech.*, **13** (1995) 1523–1541.

Shi, M.H., Wang, H. and Hao, Y.L., Experimental investigation of the heat and mass transfer in a centrifugal fluidized bed dryer, *Chem. Eng. J.*, **78** (2000) 107–113.

Shilton, N.C. and Niranjan, K., Fluidization and its applications to food processing, *Food Structure*, **12** (1993) 199–215.

Smith, P.G., Introduction to food process engineering, Kluwer Academic, New York, 2003.

Smith, P.G. and Nienow, A.W., Particle growth mechanisms in fluidised bed granulation, part I. The effect of process variables, *Chem. Eng. Sci.*, **38** (1983) 1223–1231.

Soponronnarit, S., Wetchacama, S., Trutassanawin, S. and Jariyatontivait, W., Design, testing, and optimization of vibro-fluidized bed paddy dryer, *Drying Tech.*, **19** (2001) 1891–1908.

Temple, S.J., Temple, C.M., van Boxtel, A.J.B. and Clifford, M.N., The effect of drying on black tea quality, *J. Sci. Food Agric.*, **81** (2001) 764–772.

Vanecek, V., Markvart, M. and Drbohlav, R. (trans. Landau, J.), Fluidized bed drying, Leonard Hill, London, 1966.

Vazquez, A. and Calvelo, A., Modelling of residence times in continuous fluidized bed freezers, *J. Food Sci.*, **48** (1983) 1081–1085.

White, A., Batch fluid bed drying, *Food Processing*, **52** (3) (1983) 37–39.

Chapter 5
Granulation

Granulation and particle growth

The manufacture of particles with specific properties is of increasing importance in food processing. Particle size is often, but by no means always, the most important of these properties and food particles may range in size from a few microns to several millimetres, and sometimes larger. There is no comprehensive theory which covers size enlargement techniques and therefore the operation of such processes relies heavily on empirical knowledge. There are numerous terms in use to describe powder production methods and some terms have different meanings in different industries. The word 'granulation' (Sherrington and Oliver, 1981) is probably the best all-embracing term and may be taken to mean 'the production of granules', either by producing assemblies of smaller particles or by breaking large pieces into smaller ones. Although this wide definition of granulation allows size reduction to be considered as a particle production process, this chapter is concerned only with an increase in particle size, including the addition of coatings or layers, an operation which may be carried out in a gas-solid fluidized bed. Thus granules may take a number of forms including agglomerates and layered or coated particles (Figure 5.1). In the food industry the coating of particles is also referred to as encapsulation or microencapsulation (Teunou and Poncelet, 2002).

In the context of fluidized bed granulation, the term 'initial' or 'primary' particle is used to refer specifically to particles charged to the fluidized bed, before growth takes place, and the word 'granule' is used to mean any product particle whatever its morphology; terms such as agglomeration and layering are used to describe individual modes of growth. Despite the often contradictory terminology which is used in the literature, these definitions have been adhered to in this chapter.

There are many reasons for producing powders of a specified size and structure and these may include one or more of the following:

(1) to improve the flow of the product as it leaves a packet or container
(2) to improve product appearance
(3) to improve material handling in intermediate process steps, e.g. flow in and out of process equipment
(4) to impart a particular particle structure or property to the final product.

In the case of agglomeration the new particle structure may be to aid dissolution or reconstitution when placed in water. This gives rise to the term 'instantising' (Sherrington and Oliver, 1981). Rapid dissolution favours small particles because of their high specific surface but these may not be wetted readily when placed in water. An agglomerated structure (Figure 5.1a) allows the larger particle to sink and disperse more readily in water. Water is then taken up into the void spaces

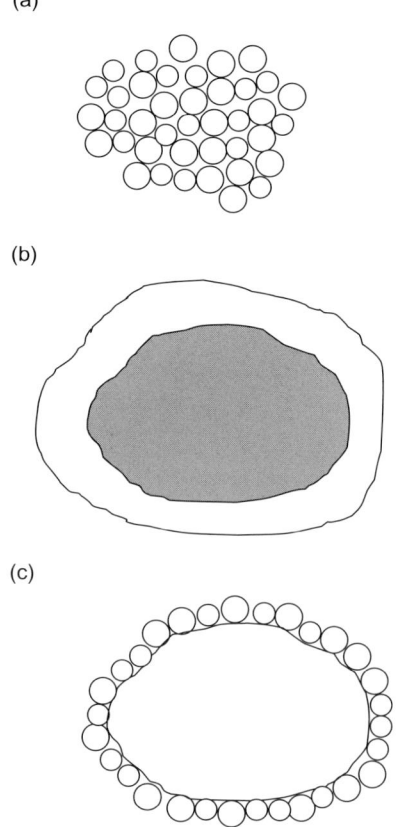

(a)

(b)

(c)

Figure 5.1 Granule morphology. From Smith, P.G., Introduction to food process engineering, Kluwer, 2003, figure 13.19. With kind permission of Springer Science and Business Media.

of the agglomerate, the solid bridges dissolve and rapid dissolution of the primary particles then follows. Agglomerated products include instant milk powder, starch, coffee, instant soups, dry pudding mixes, cocoa and sugar (Smith, 2003). In the case of coating or encapsulation, the purpose may be to increase shelf-life or control the release of constituents within the particle (Dewettinck and Huyghebaert, 1998) by providing a coating which dissolves at a given rate in a particular environment, for example at a particular temperature or pH.

Excluding size reduction techniques, granulation processes may be classified in three major groups. In drop formation methods a melt (e.g. in prilling, spray cooling or globulation) or a solution (e.g. in spray drying) is broken up into small droplets. Spray drying requires the addition of thermal energy whereas in the other processes the melt solidifies in an air stream or is cooled on a continuously moving belt. In this context it is important that spray drying should be seen as a particle formation process. The second group includes compaction and extrusion processes in which powders (sometimes containing a wide distribution of particle sizes) or soft solids are forced together under great pressure and formed into pre-determined shapes. Often a liquid binder is included in the powder feed.

The most important group of granulation processes may be termed agitation methods, including fluidized bed and spouted bed granulation, in which a liquid binder is added to particles being agitated in some kind of mixing device. In food systems the binder is often an aqueous solution of, for example, lactose or dextrose, gelatine or a food gum (Smith, 2003); celluloses, starch derivatives or polyethylene glycol are also used (Härkönen *et al.*, 1993). Alternatively the binder may be a molten fat (Sommer *et al.*, 2002). Thus it is possible to use any powder mixer (fluidized bed, inclined rotating pan, rolling drum or spouted bed) as a granulator. The solids being granulated may be a mixture of powders which are held together by the binder or may consist of a single component. The binder will usually undergo a change of phase: either the solidification of a melt or the removal of a solvent leaving behind the solute as the 'adhesive'. In the latter case it is usually necessary to have a drying stage; one of the significant advantages of fluidized bed granulation is that both particle growth and drying can be achieved in the same equipment. Very fine powders can be agglomerated without binder but the product granules are weak and the final size is limited. In many food applications the particle surface is wetted with a fine spray of water or by the injection of steam, which promotes the dissolution of the particle surface and allows particles to fuse together. A major example of the use of this technique is the production of instant milk powders following the spray drying of milk (Masters, 1983).

Particle-particle bonding

An understanding of the nature of the bonds between particles in an agglomerate is fundamental to an understanding of granulation. The strength of interparticle bonds determines not only granule strength (which is itself an important property) but also granule size.

Bonding mechanisms

The effect of different bonding mechanisms, and of the primary particle size on granule strength, is shown in Figure 5.2. Four mechanisms (Capes, 1979; Sherrington and Oliver, 1981) need to be considered. First, intermolecular forces. The attractive force between particles is inversely

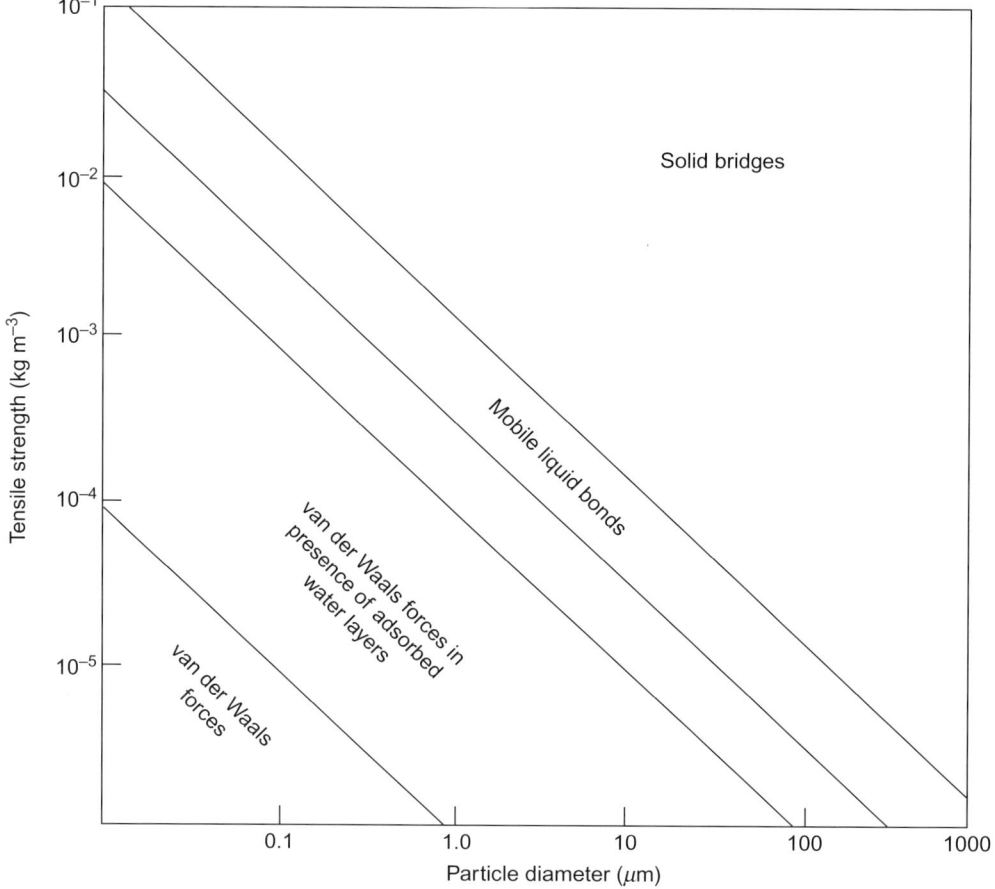

Figure 5.2 Tensile strength of agglomerates. From Smith, P.G., Introduction to food process engineering, Kluwer, 2003, figure 13.22. With kind permission of Springer Science and Business Media.

proportional to the seventh power of the separation distance of the particles. Surface roughness increases the effective separation distance and consequently van der Waals forces are not usually significant in granulation systems. Second, for dry particles, electrostatic forces can be of the same order of magnitude as van der Waals forces. Third, liquid bonds. Adsorbed liquid layers on the particle surface have the effect of smoothing out surface roughness and decreasing particle separation distances. Consequently, van der Waals forces may increase in magnitude significantly. Of greater significance are the mobile liquid bonds which are present in a granule initially when the solids to be granulated are contacted with liquid binder. They are usually a prelude to the formation of permanent solid bridges. Increasing the quantity of liquid changes the nature of the bonds and influences the overall granule strength. An assembly of particles containing bonds at the contact point between individual particles is described as being in the pendular state (Newitt and Conway-Jones, 1958) (Figure 5.3a). Increasing the liquid content of the agglomerate gives rise to the funicular state (Figure 5.3b) and finally the capillary state (Figure 5.3c) in which the interparticle voidage is saturated with liquid. This is amenable to theoretical analysis; the tensile strength of a capillary state granule is given by

$$\sigma = \frac{8(1-\varepsilon)\gamma}{\varepsilon d} \qquad 5.1$$

where ε is the interparticle voidage within the granule, γ is the surface tension of the liquid and d is the diameter of the constituent particles (Capes, 1979). Finally, solid bridges. Whilst the analysis of the strength

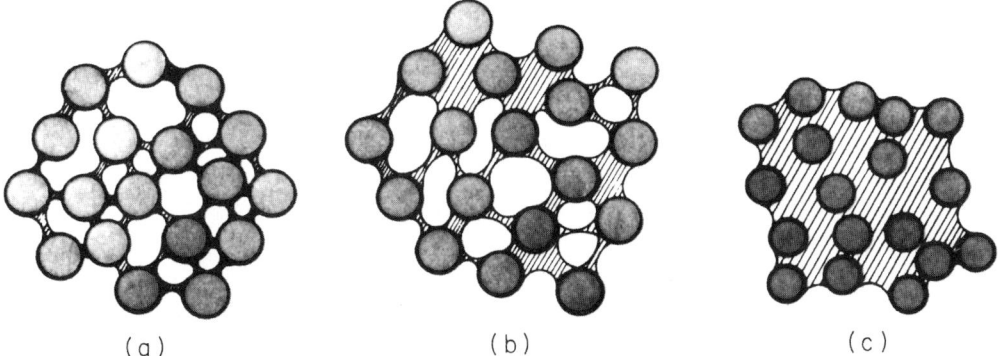

(a) (b) (c)

Figure 5.3 Mobile liquid bonds: (a) pendular; (b) funicular; (c) capillary. From Sherrington, P.J. and Oliver, R., Granulation, Heyden, 1981. Copyright John Wiley & Sons Limited. Reproduced with permission.

of moist agglomerates and particles bound by liquid is well developed (Rumpf and Schubert, 1974), solid bridges between particles do not lend themselves readily to theoretical treatment. The strength of crystalline bridges depends not only on the amount of material present but also upon its structure (Pietsch, 1969a). A finer crystal structure results in stronger bonds and there is some correlation between bond strength and higher drying temperatures. By assuming that all the material available for forming solid bridges is distributed uniformly over all points of contact between constituent particles in the granule and that the material has a constant tensile strength, the strength of an agglomerate can be defined (Pietsch, 1969b) by

$$\sigma = \varepsilon\theta f \qquad\qquad 5.2$$

in which θ is the intrinsic tensile strength of the bridge and f is the fraction of the void volume filled with binder. Little more can be said from a theoretical point of view. For a given concentration of binder, particle size and granule size, granule strength is a function of the structure and physical properties of the binder used and further information can only be obtained by experiment. However, experimental measurements of granule strength are difficult and somewhat tedious. Further, it is difficult to measure the strength of the solid bridge independently because it cannot easily be cast into a form amenable to standard tests.

Growth mechanisms in granulation

In conventional agglomeration processes granules grow by the successive addition of primary particles to an agglomerate. This occurs when two particles or two granules are brought into contact with sufficient liquid binder present to hold the two species together. In tumbling beds of powder (e.g. the pan granulator), two agglomerates colliding will be kneaded together by the tumbling action of the mixer and, because of their surface plasticity, form an approximately spherical granule. This sequence of events, forming a nucleation stage, continues until the granules are sufficiently large that the torque tending to separate them is too great to allow a permanent bond (Capes and Danckwerts, 1965). Subsequent growth occurs by a 'crushing and layering' mechanism (Capes, 1979; Sherrington and Oliver, 1981) in which the smallest and weakest granules are crushed by larger ones and the material redistributed around the surface of the large granule in a uniform layer (Figure 5.4). Note that a distinction must be drawn here between the layering of smaller particles around a larger granule and the deposition of solute which provides a coating or layer around a

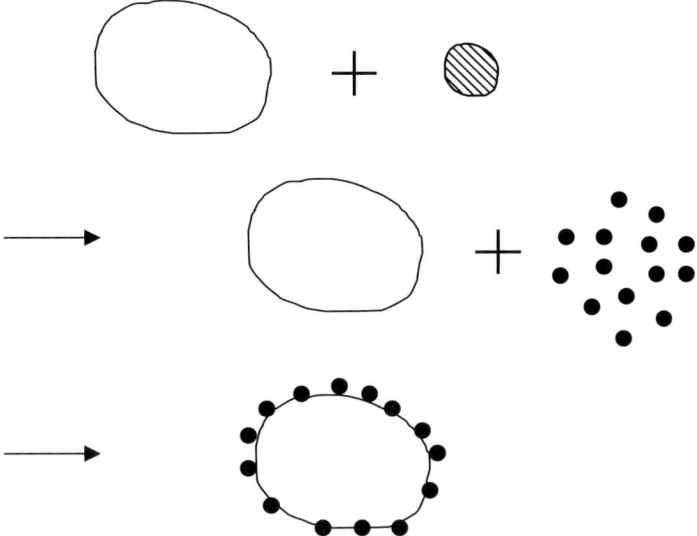

Figure 5.4 Crushing and layering mechanism.

primary particle. The latter is one of the major mechanisms of particle growth in fluidized bed granulation.

In contrast, it has been claimed that coalescence is mainly responsible for growth (Kapur and Fuerstenau, 1966) although Linkson *et al.* (1973) showed that this was due to the use of a wide size distribution of particles which form strong granules, resistant to crushing, which then grow by coalescence until a terminal size is reached. Some of this work (Newitt and Conway-Jones, 1958; Capes and Danckwerts, 1965) showed that the amount of liquid required for granulation was equal to the saturation content of the voids and therefore a function of the packing density of the original particles. Sherrington (1968) found that only one half of this amount was required and on this basis developed a model to relate liquid phase content and average granule size (see Agglomeration model, below) which was subsequently adapted by Smith and Nienow (1983b).

Fluidized bed granulation

Introduction

Despite the importance of fluidized bed granulation within the food industry, there is relatively little published literature on the operation of fluidized beds used for agglomeration, instantising, coating or encapsulation of food materials and consequently very little fundamental information is available about the prevailing mechanisms of granulation within food systems. However, the technique is very widely used in other industries and, as Shilton and Niranjan (1993) explain, much useful information can be obtained from the extensive literature on these applications of granulation which extends back over several decades. Only Ormos and co-workers (Ormos, 1973; Ormos *et al.*, 1973a–c) and Smith and Nienow (Smith and Nienow, 1981, 1982, 1983a,b) have conducted systematic investigations into the effects of process variables on growth and growth mechanisms.

Principles of operation of fluidized bed granulation

The term 'fluidized bed granulation' (Figure 5.5) refers to processes which produce granules or dry powder from a solution or slurry in a fluidized bed to which sensible heat is applied. Growth of bed particles, creation of new particles and drying of the product may all take place. Heat for evaporation of the solvent or for the removal of moisture from bed particles can be supplied either in the fluidizing air or through the bed walls, and the wet feed material may be introduced under, or sprayed onto, the bed surface. An excess of liquid feed, either over the whole bed or in a localised region, produces excessive and uncontrollable particle agglomeration and leads to a loss of fluidization or what Nienow and Rowe (1975) called 'wet quenching'. Smith and Nienow (1981) called the defluidization phenomenon, which results in the failure of the process, simply 'bed quenching' and used the term 'dry quenching' when defluidization was the result of the excessive formation of dry granular material and reserved 'wet quenching' for cases where failure is caused by excessive free liquid.

For a given bed outlet temperature, the liquid feed rate must not exceed that which will saturate the outlet air stream. If this condition is not obeyed, the bed material will become increasingly over-wet. Continued operation under these conditions will lead rapidly to wet quenching and the failure of the process (Dewettinck *et al.*, 1998).

Despite the apparent incompatibility of free liquid and fluidized solids, the use of a fluidized bed for granulation offers several advantages over more traditional methods such as spray drying or pan granulation. Good heat transfer, uniform bed temperatures and close temperature control are advantages which are particularly important when heat-sensitive materials are being handled. In comparison with a spray drier, a fluidized bed represents a large reduction in plant volume for the same throughput. Closer control of the physical properties of the product, such as particle size, flow characteristics and bulk density, is possible; a fluidized bed relies not only on the fine atomisation of the feed liquid but also its interaction with existing bed

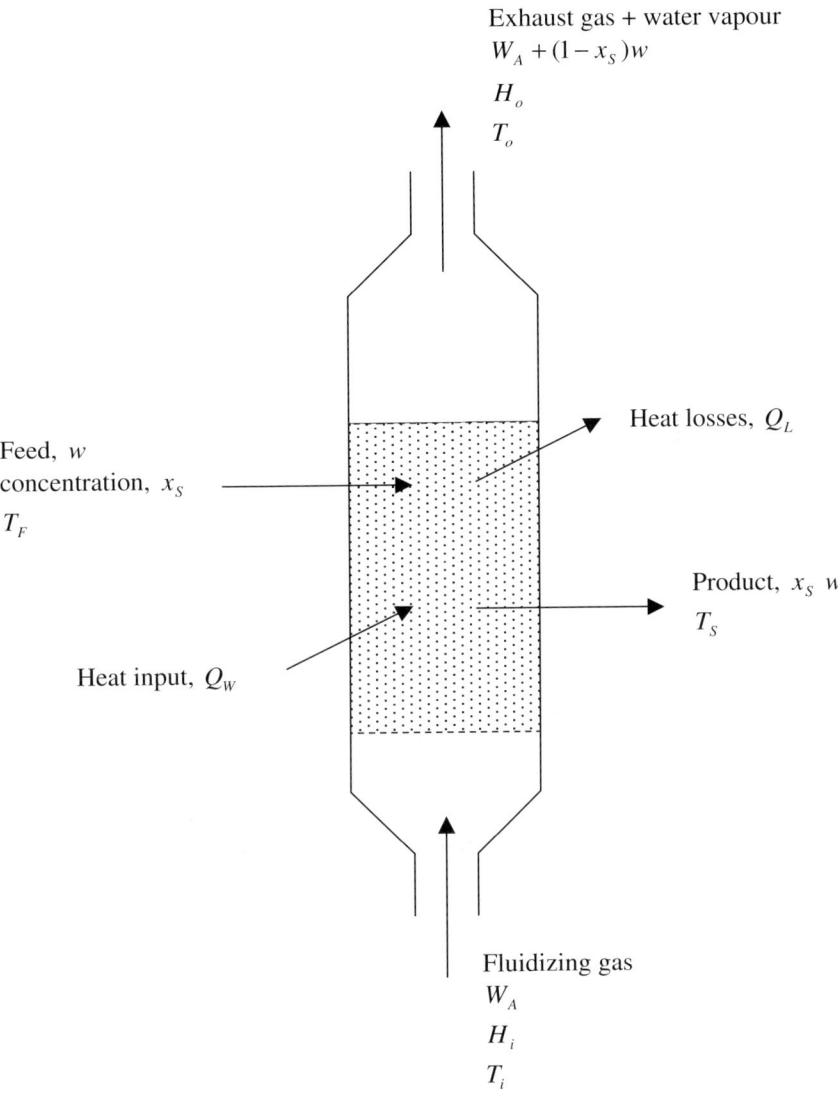

Exhaust gas + water vapour
$W_A + (1 - x_S)w$
H_o
T_o

Heat losses, Q_L

Feed, w
concentration, x_S
T_F

Product, x_S w
T_S

Heat input, Q_W

Fluidizing gas
W_A
H_i
T_i

Figure 5.5 Schematic diagram of a fluidized bed granulator.

particles. Thus it is possible to form particles in a fluidized bed of a size larger by an order of magnitude (Nienow and Rowe, 1975).

A fluidized bed is normally characterised by the absence of temperature gradients because of the inherent rate of particle mixing. However, Smith and Nienow (1982) measured temperature gradients within the bed both when segregation of large agglomerates occurred and during stable particle growth, in the region immediately below the spray nozzle which was itself immersed just below the bed surface. An example of an isotherm constructed from steady-state temperature

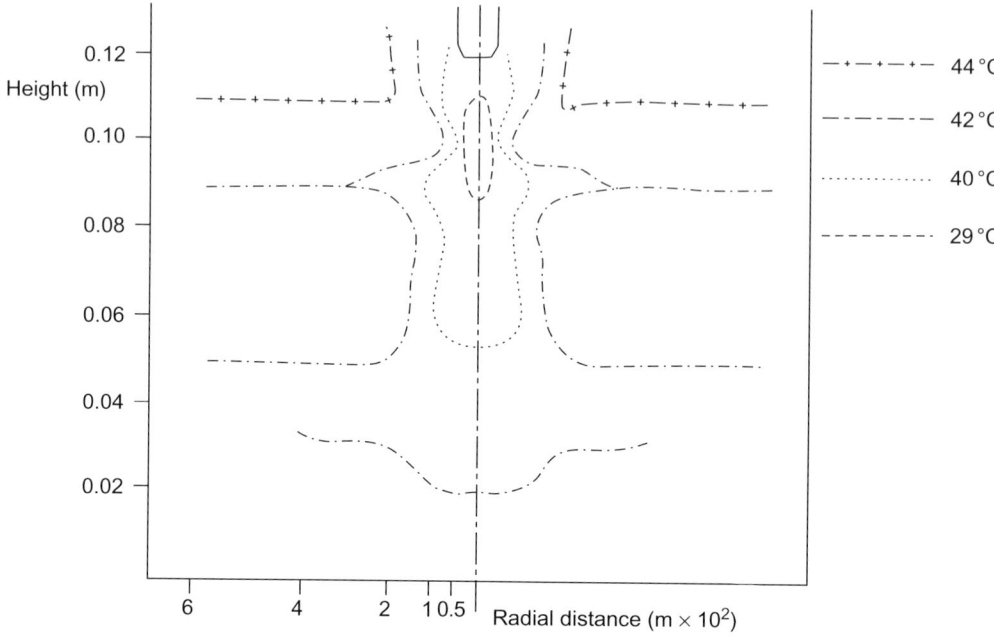

Figure 5.6   Temperature gradients in fluidized bed granulation (Smith, 1980).

measurements in the plane of a bed diameter is shown in Figure 5.6; the low-temperature region increased in size with increasing liquid feed rate.

Solvent in the feed liquid is evaporated in this well-defined zone close to the spray nozzle, and from the surface of the bed particles with which it inevitably comes into contact, resulting in agglomeration. The extent to which clumps of agglomerated particles remain intact determines the outcome of the fluidized bed granulation process. Bed quenching results if insufficient breakdown takes place, and breakdown into smaller agglomerates, to an equilibrium size, will give a product of agglomerated granules (Smith and Nienow, 1983b). Further reduction and tearing apart of smaller agglomerates ultimately produces a single bed particle with associated binder; in other words, a layered or coated granule. Thus an increase in the excess gas velocity causes a bed to move from quenching through agglomeration to layered growth (Smith and Nienow, 1983a). Smaller bed particles are more likely to form permanent bonds, and to quench, because of their smaller inertia. With larger initial particles, the mean diameter of product granules decreases and it is possible to achieve layered growth under conditions which would otherwise lead to quenching.

Material and energy balances

The fundamental material and thermal energy balance equations have been set out by Scott *et al.* (1964) and by Nienow and Rowe (1975) amongst others. Referring to Figure 5.5, the mass balance is

$$\text{air in} + \text{liquid in} = \text{air out} + \text{vapour out} + \text{solids out} \qquad 5.3$$

or

$$W_A + w = W_A + (1 - x_S)w + x_S w \qquad 5.4$$

where x_s is the mass fraction of solute in the feed liquid. Consequently the energy balance is

$$
\begin{aligned}
& W_A c_A (T_i - T_o) + w c_F (T_F - 0) + Q_W \\
& = (1 - x_S) w [c_V (T_o - 0) + h_{fg}] + x_S w (T_S - 0) + Q_L
\end{aligned} \qquad 5.5
$$

where respective terms represent: the enthalpy change of the fluidizing gas; enthalpy of the feed stream; possible heat supplied through the bed wall Q_W; the enthalpy of the vapour in the exhaust gas (including the latent heat of vaporisation h_{fg}); enthalpy of the solid product stream; and heat losses from the system Q_L. The nature of heat transfer in a fluidized bed means that the product outlet temperature T_S is very close to the exhaust gas temperature T_o.

A moisture balance may also be written

$$(1 - x_S)w = W_A (H_o - H_i) \qquad 5.6$$

where H_i and H_o are the absolute humidities of the fluidizing gas and exhaust gas respectively.

In these equations it is assumed that the solid product has an acceptable moisture content (which could be zero). Equation 5.6 stipulates that for a given bed outlet temperature, the liquid feed rate must not exceed that which will saturate the outlet air stream. Scott *et al.* (1964) pointed out that, if equation 5.6 is not obeyed, the bed material will become increasingly over-wet. Continued operation under these conditions will rapidly lead to wet quenching and the failure of the process. Nienow and Rowe (1975) used the energy balance (equation 5.5) to illustrate a possible fundamental difference between fluid bed granulation and spray drying. For a spray drier $Q_W = 0$ but with a fluidized bed a considerable amount of the required heat may be put in through the bed walls. The term Q_W must be substantial if the fluidizing gas flow rate is to be reduced significantly (Vance and Lang, 1970).

Batch and continuous operation: population balance

It is possible to operate a fluidized bed granulator in either a batch or a continuous mode. Batch operation produces a continuous increase in bed mass and therefore, if attrition and particle breakdown effects are not dominant, a continuous increase in bed particle size. This necessitates a gradual increase in the volumetric air flow through the bed to compensate for the increasing minimum fluidizing velocity and thus maintain the 'degree' or 'quality' of fluidization (Ormos *et al.*, 1973a). With continuous operation it is desirable to maintain a stable particle size distribution; in other words, to satisfy the population balance (Randolph and Larson, 1971). Clearly, in order that granules do not grow to be too large, seed particles or nuclei must be added to the bed together with the removal of large particles. Metheney and Vance (1962) controlled the particle size distribution by adjusting the size of seed particles and the liquid to solid feed ratio and by means of an in-bed classification device. In some cases particle size reduction has been achieved inside the fluidized bed by grinding with high-velocity air jets (Barsukov and Soskind, 1973; Bjorklund and Offutt, 1973). More recently, Tsujimoto *et al.* (1998) reported the use of an opposed pulsed air jet assembly to produce particles smaller than $100\,\mu$m with better solubility and higher compressibility compared with those produced by conventional fluidized bed granulation.

In the case of fluidized bed coating (i.e. the production of layered granules) continuous processing requires a continuous feed of both initial particles and binder in proportion so as to give the desired layer thickness. Teunou and Poncelet (2002) point out that only a continuous process for bulk food applications of coating is likely to be economically viable. Rümpler and Jacob (1998) describe a fluidized bed system manufactured by Glatt which is designed for the continuous cooling, drying, agglomeration, granulation and coating of powdered and granular foods. The design allows different gas velocities and inlet gas temperatures (between 20°C and 250°C) to be used at various points along the length of the bed. Liquid binder is introduced using multiple nozzles in either top or bottom spray mode. The equipment is able to handle particle sizes between $100\,\mu$m and 3 mm. Some details of the variation in velocity and temperature in different bed sections are given by Rümpler and Jacob (1999) together with information about the filters for the separation of elutriated particles and exhaust gas. The feed of inlet particles, and the discharge of product granules, is achieved by means of a cellular wheel impeller which seals the fluidized bed from the external environment.

Bed quenching

Nienow and Rowe (1975) suggested that the capacity of a fluidized bed granulator is limited by the amount of free liquid that can be tolerated in the bed. Certainly, defluidization due to bed quenching is one of the major reasons for unsuccessful operation of fluidized bed granulators; a proportion of the literature reports such problems, although whether wet or dry quenching is the cause is not always clear. Very early in experimental studies it was realised that good liquid distribution would prevent bed quenching and serious caking problems and consequently atomising spray nozzles were used (Jonke *et al.*, 1957) with the idea of reducing the amount of liquid feed associated with each bed particle. Rapid particle mixing will prevent the build-up of localised moisture and it has been suggested (Nienow and Rowe, 1975) that the mixing obtained in a fluidized bed, being a good approximation to perfect mixing, combined with top-spraying of feed, enables granulation to be carried out without bed quenching. Further, a much more ordered particle circulation pattern, as in for example a spouted bed (Mathur and Epstein, 1974) (see Spouted bed granulation, below), is likely to prevent agglomeration and hence quenching. In most of the reported work in which a liquid feed is introduced into a fluidized bed, two-fluid atomising nozzles have been employed, either entering through the bed wall and below the bed surface or positioned in the freeboard region with feed being sprayed onto the fluidized surface. Detailed discussion of the selection of nozzles for fluidized bed applications is given by Legler (1967) who concluded that two-fluid nozzles are the most satisfactory. However, the literature contains several references to severe problems encountered with this form of liquid injection despite its widespread use. These include caking of the nozzle (Hawthorn *et al.*, 1960), bed walls and distributor plate (Rankell *et al.*, 1964), nozzle blockage (Fukomoto *et al.*, 1970), nozzle erosion (Legler, 1967) and severe agglomeration or quenching of the fluidized solids (Bjorklund and Offutt, 1973).

Several workers (Hawthorn *et al.*, 1960; Metheney and Vance, 1962; Bjorklund and Offutt, 1973) have varied the atomising nozzle geometry and position in an attempt to improve performance or to eliminate caking problems. Jonke *et al.* (1957) reported that positioning the nozzle in the freeboard and spraying liquid feed onto the fluidized surface results in caking of feed material both on the nozzle and in the bed. There is also a danger of overspraying onto the bed walls with this arrangement (Metheney and Vance, 1962). Nozzle caking was still a problem when the nozzle was mounted in a hole cut in the distributor plate with the spray directed upwards (Jonke *et al.*, 1957). In a study of pharmaceutical granulation, Davies and Gloor (1971) found that the number of large agglomerates formed in the bed increased as the atomising nozzle was lowered towards the bed surface.

In pharmaceutical granulation the nozzle has usually been located in the freeboard of the bed. This is true also of the majority of the work reported on coating of food particulates where it is referred to as 'top spraying' (Eichler, 1996; Teunou and Poncelet, 2002) although other nozzle positions have been investigated (Teunou and Poncelet, 2002). When side entry into the bed is used the vertical position within the fluidized layer has variously been claimed to be either of only minor importance (Jonke *et al.*, 1957) or to be critical in preventing bed quenching (Hawthorn *et al.*, 1960; Bjorklund and Offutt, 1973). A detailed study of problems encountered with this nozzle geometry is reported by Otero and Garcia (1970) who present expressions to describe the extent of formation of lumps and cakes of feed material as a function of operating variables.

The degree of atomisation of the feed may also have considerable effect on bed quenching. Feeding sodium sulphate solutions through a hypodermic needle (Bakhshi and Chai, 1969), and thus without atomisation, resulted in large agglomerates which segregated and formed a defluidized layer on the distributor. At the other extreme, Ormos *et al.* (1973c) found that increasing the flow of atomising air beyond a certain point caused liquid to penetrate too deeply into the bed and clog the distributor plate. It is possible to use mechanical methods of preventing bed quenching. For example, Watano *et al.* (1998) used a rotating blade at the bottom of the fluidized bed to reduce segregation which they claim produced spherical and well-compacted granules. A new development is a conical fluidized bed equipped with a vortex orifice air distributor. This consists of four tangential injection nozzles 40 mm in diameter along the inner wall of the bed and a two-fluid nozzle placed at the centre of the distributor (Kurita and Sekiguchi, 2000). The swirling air flow which is generated is reported to be responsible for a centrifugal and buoyant effect on the bed granules and good control of particle size.

In an early attempt to remove mists from a gas stream, by using a fluidized bed as a kind of filter (Meissner and Mickley, 1949), it was found that the operation worked well if the bed particles had a porous structure, and that when non-porous particles were used fluidization ceased (i.e. the bed quenched) at very low moisture contents. This view is supported in work on the effect of bed moisture on the fluidization characteristics of fine powders (D'Amore *et al.*, 1979) in which it was shown that porous materials can tolerate considerably more liquid than non-porous particles (such as glass ballotini, sand and limestone) before what these workers call 'bed compaction' occurs. Smith and

Nienow (1983a) observed a delay in the start of particle growth when binder was added to a bed of porous particles and stable fluidization under conditions which produced quenching with non-porous particles. Nitrogen adsorption measurements showed that the pore surface area of alumina decreased as spraying proceeded, indicating that an effective reduction in pore volume was taking place.

Of the other variables mentioned which determine whether fluidized beds quench or operate in a stable condition, fluidizing gas velocity would seem to be the most important. The superficial gas velocities required to give sufficient mixing to avoid caking or lump formation in the bed have been quoted for particular systems (Buckham *et al.*, 1964) although no indication of the relative gas velocities involved are given. Gluckman *et al.* (1976) found that the defluidization velocity was directly proportional to the amount of liquid introduced; in other words, at a higher velocity more liquid was required to produce bed quenching. Smith and Nienow (1983a) found that whether a bed quenches or whether stable operation can be maintained depends upon the excess gas velocity, to which particle movement in the bed is proportional. Increasing the excess gas velocity allowed granulation to continue for longer before quenching occurred. It has been observed that the chances of bed quenching are greater at high liquid feed rates (Jonke *et al.*, 1957; Crooks and Schade, 1978). Increasing the binder concentration results in faster growth and a more rapid onset of quenching; granulation can be continued for increasingly longer periods by reducing the feed concentration (Smith and Nienow, 1983a). The granulation of pharmaceutical powders at room temperature also failed due to 'overwetting' of the bed material (Davies and Gloor, 1971).

Effect of variables on growth
Rate and volume of feed

The growth rates of particles in a fluidized bed granulator increase when increasing quantities of solute or binder are introduced into the bed, by increasing either the feed rate of solution or the concentration of dissolved solids (Jonke *et al.*, 1957) and, as might be expected in a batch operation, the mean particle diameter increases with the volume of liquid introduced (Ormos *et al.*, 1973a; Crooks and Schade, 1978; Yu *et al.*, 1999). The growth data of Rankell *et al.* (1964) pass through a maximum, suggesting that a spray rate exists at which the agglomeration of bed particles is balanced by attrition and breakdown, a tendency noted in another agglomerating system (Ormos *et al.*, 1973a). Most of the information on particle growth as a function of the rate, volume and concentration of feed liquid is to be found in the published work on pharmaceutical granulation, in which agglomeration is the dominant growth mode. There is only limited information on food systems (Härkönen *et al.*, 1993).

Several authors (Davies and Gloor, 1971; Ormos *et al.*, 1973a; Crooks and Schade, 1978) have found that the rate of spraying a fixed amount of feed into the bed affects particle growth. Generally, increased addition rates (i.e. shorter total spraying times) have produced larger mean particle diameters and Davies and Gloor (1971) attribute this, as well as a slight increase in granule porosity (or lower packing density), to greater penetration of the bed by the liquid feed. However, the results described by Ormos and co-workers (Ormos *et al.*, 1973a, b) are not in agreement and show a slight decrease in mean diameter at higher rates, although no change was observed in the spread of the particle size distribution.

Nozzle position and atomising air rate

In addition to its effect on bed quenching, the position of the atomising nozzle appears to have some consequences for particle growth, particularly when 'top spraying' of the feed is employed and spray drying occurs in the freeboard of the bed (Legler, 1967; Dewettinck and Huyghebaert, 1998). Dewettinck and Huyghebaert (1998) used a model system of various protein concentrates sprayed onto sodium chloride crystals and suggested that nozzle atomisation pressure was a very significant variable; higher pressures created smaller droplets which were more likely to evaporate in the freeboard with solute becoming unavailable for layering. In addition, the increased nozzle atomisation pressure caused a drop in bed temperature as well as influencing the nature of the interaction between bed particles and droplets. Smaller mean particle diameters, due to an increased spray-drying effect, are reported when the nozzle is placed at increasing distances from the bed surface (Davies and Gloor, 1971), maximum growth occurring when the nozzle is actually below the bed surface (Rankell *et al.*, 1964). The findings of Ormos *et al.* (1973c) do not agree here; they report no change in particle size with nozzle position and claim that spray drying is a stronger function of air temperature and liquid droplet size. An equation is presented which predicts the optimum nozzle height by avoiding overspray onto the bed walls. Increasing the air-to-liquid ratio (normalised air ratio or NAR) through the nozzle gives a smaller particle size which has been attributed to both attrition (Buckham *et al.*, 1964) and to the production of finer liquid droplets (Ormos *et al.*, 1973c; Yu *et al.*, 1999).

Bed temperature

It has been verified experimentally (Rankell *et al.*, 1964) that allowable liquid flow rates are directly proportional to the air inlet temperature; however, the true effect of bed temperature is unclear. Granulation of pharmaceutical powders with aqueous binding solutions (Davies and Gloor, 1971; Crooks and Schade, 1978) below 100°C has shown that mean particle size decreases with increasing bed temperature. Davies and Gloor (1971) increased the air inlet temperature to the bed over the range 25–55°C (giving a higher bed temperature for the same liquid flow rate) and claimed this to be responsible for decreased penetration and wetting of the fluidized solids, and consequently a decrease in mean particle diameter. Conversely, Jonke *et al.* (1957) proposed that higher temperatures result in a more rapid evaporation on the particle surface – before penetration of the intraparticle pores is possible – and therefore fracture of the particle, due to vaporisation within the pores, is avoided. In this way the net growth is greater at higher temperatures. The optimum bed temperature for fluidized bed coating is discussed in Fluidizing gas velocity and particle mixing, on page 164.

Fluidizing gas velocity

For a given fluidized bed geometry and particle size, the superficial gas velocity through the bed is the most important and fundamental variable; it affects bed expansion, the extent of bubbling and particle mixing. For this reason it is very surprising that the effect of velocity on the granulation process, and particularly on particle size, remains largely uninvestigated. Very little information is available and the majority of experimental studies have been carried out at a constant gas velocity chosen, for example, to give the least elutriation and maximum cyclone efficiency (Rankell *et al.*, 1964) or perhaps the lowest possible velocity consistent with adequate fluidization (Bjorklund and Offutt, 1973). Qualitative observations have suggested that an increase in gas velocity leads to less agglomeration due to the higher degree of particle-particle impact and attrition, and that a more uniform particle size distribution is produced with higher fluidizing air rates (Bakhshi and Chai, 1969). Ormos *et al.* (1973c) present data which show that an increase in bed expansion produces a linear decrease in mean particle diameter because of increased abrasion, although the corresponding gas velocities are not given.

Only Smith and Nienow (1983a) have reported a systematic investigation of the effects of gas velocity on granule size and on particle growth mechanisms. These authors used model materials (non-porous glass powder or porous alumina) and solutions of either benzoic acid or polyethylene glycol 4000 in a series of batch experiments. In each case the binder was dissolved in methanol in order to lower the evaporation temperature and thus permit the observation of particle growth through a glass-walled bed in addition to the removal of samples for size analysis. With a weakly agglomerating system (using benzoic acid as binder) particles grew by agglomeration at low excess gas velocities and by layering at high velocities. In a strongly agglomerating system (with polyethylene glycol as binder) no layered granules were observed in the studied velocity range but increasing the excess gas velocity produced lower growth rates.

Smith (1980) and Smith and Nienow (1983a), using data from the published literature, recalculated reported superficial gas velocities in terms of the excess gas velocity (Table 5.1) and noted the relationship between $u - u_{mf}$ and the reported mode of particle growth. Those systems falling into category (1), higher excess velocities, were reported to exhibit layered growth; those in category (2) both layered growth and growth by agglomeration; and those in category (3), lower velocities, either grew by agglomeration alone or were reported to quench almost immediately. More recently, Yu *et al.* (1999) found that sodium benzoate particles granulated with sodium benzoate solution (the solvent was not specified) grew mainly by layering, although reducing the excess gas velocity changed the dominant growth mechanism to agglomeration.

Table 5.1　Correlation of reported growth mechanism with excess gas velocity.

Reference	Superficial gas velocity u (m s^{-1})	Relative gas velocity $\dfrac{u}{u_{mf}}$	Excess gas velocity $u - u_{mf}$ (m s^{-1})
(1) Layered growth			
Fukomoto *et al.* (1970)	0.4–0.9	3–8	0.27–0.79
Lee *et al.* (1962)	0.4–0.6	4	0.30–0.45
Grimmett (1964)	0.4	9	0.35
Buckham *et al.* (1966)	0.3	4	0.23
Otero and Garcia (1970)	0.15–0.47	5–15	0.12–0.44
(2) Layered growth and agglomeration			
Jonke *et al.* (1957)	0.4–0.6	4–6	0.30–0.50
Bakhshi and Nihilani (1973)	0.15	10–17	0.13
(3) Agglomeration or quenching			
Bjorklund and Offutt (1973)	0.10–0.15	2.5	0.06–0.09
Frantz (1958)	0.13–0.27	2–4	0.06–0.20
Bakhshi and Chai (1969)	0.10–0.20	7–12	0.09–0.18

Particle size

Size-dependent growth of particles has been reported in a few cases (Lee *et al.*, 1962; Grimmett, 1964; Fukomoto *et al.*, 1970; Crooks and Schade, 1978). Grimmett (1964) suggested that larger particles remain for longer in the spray zone because the atomising air forms a barrier through which large particles selectively penetrate. Smith and Nienow (1983a) found that increasing the size of initial particles with a weakly agglomerating model system produced a similar effect on growth mechanism to increasing the excess gas velocity. With particles originally of mean diameter 270 μm, quenching occurred within 170 minutes, whilst with larger particles (mean size 437 μm) at the same excess gas velocity, granulation continued for 480 minutes. In the first case the granules grew by agglomeration, in the second case by layering. Particle growth rates, relative to the initial particle size, were significantly less for larger particles than for the smaller particles with both weakly and strongly agglomerating systems. In all these experiments the excess gas velocity was kept constant by continually adjusting the superficial gas velocity to compensate for the increase in minimum fluidizing velocity of growing particles.

Binder properties

The effect of different binders and solutes and their physical properties on the granulation process and on the nature of the product granules can be judged both from the pharmaceutical granulation literature and from recent work with food systems. Higher concentrations of feed solution (at a given solution flow rate) obviously increase the amount of solid material available to produce growth of bed particles, but it has also been clearly demonstrated that different binders (at the same concentration) have very different growth characteristics (Davies and Gloor, 1972). Water alone is a very poor granulating agent (Rankell *et al.*, 1964; Davies and Gloor, 1973) while diluted syrup gave granules which were too small for tabletting and only an aqueous gelatine solution produced a satisfactory granulation (Rankell *et al.*, 1964).

The literature contains relatively little data on the physical properties of binder solutions and therefore conclusions must be drawn largely from qualitative observations. For example, Davies and Gloor (1972) have linked the effectiveness of binders with their adhesiveness or tackiness and have found that more viscous binder solutions increase the size of granules and reduce the bulk density. Hydroxypropylcellulose (HPC), with which atomising difficulties were encountered because of its viscosity, was responsible for the largest increase in mean particle size. This was closely followed, in its effect on particle growth, by aqueous gelatine solutions the viscosity of which increases exponentially with concentration. HPC solutions increased mean particle diameters to 257 μm at 2% w/w and to 406 μm at 4.25% w/w, whilst solutions of povidone produced granules with mean diameters of only 200 μm and 250 μm respectively. This growth effect is reflected in the granule friability data which showed that gelatine and HPC solutions produced significantly stronger granules than other binders. Gelatine has also been used to granulate quartz sand to produce particles of high wear resistance (Ormos *et al.*, 1973a).

Crooks and Schade (1978) successfully granulated lactose particles with aqueous solutions of polyvinylpyrolidone (PVP). Wurster (1960) reports using a variety of binders in preparing coated drug particles. Solutions of carbowax (polyethylene glycol), 'simple syrup', starch and combinations of these in both water and methanol have been used, but no information is supplied about their relative performance. Yu *et al.* (1999) observed layered growth when sodium benzoate particles were granulated with sodium benzoate solution (the solvent is not specified) whereas the granulation of sand with a solution of sodium carboxymethylcellulose occurred mainly by agglomeration.

Two studies have attempted to relate binder behaviour to physical properties. Smith and Nienow (1983a) observed layered growth when granulating glass powder with 10% benzoic acid but agglomeration became the dominant growth mechanism when 5% polyethylene glycol 4000 was substituted. Visual observation of the granule morphology was confirmed by good agreement between experimental growth curves and the growth models set out in the next section. Further, although the surface tension of the two solutions was similar, the viscosity of polyethylene glycol 4000 solution increased rapidly as the solvent was evaporated and therefore was less able to flow around a particle to cover the surface. Consequently layered growth was less probable than with the less viscous benzoic acid solution. In contrast to these conclusions about the effect of binder viscosity, Dewettinck *et al.* (1998) concluded that the wet film and not the solution characteristics has the greatest influence on growth mechanisms. These workers used smooth spherical glass beads (365 μm in diameter) as a model system to investigate the agglomeration tendency of several gums (locust bean gum, carboxymethylcellulose (CMC), sodium alginate and kappa-carrageenan) during top-spray fluidized bead coating. They found no difference between high and low viscosity CMC binder solutions and high viscosity locust bean gum in terms of their tendency to produce agglomeration. Scanning electron micrographs showed that high-viscosity carboxymethylcellulose did not spread out over the particle surface and that high-viscosity, yet highly hygroscopic, locust bean gum gave a smooth coating. They concluded that the poor coating ability of CMC was due to its higher hygroscopicity.

Fluidized bed granulation growth models

Layered growth model

When layered, or 'onion-ring', growth takes place in a batch granulator, a simple expression for the increase in mean particle diameter with time can be obtained by assuming the uniform distribution of binder around an idealised core particle (Figure 5.7). Suppose the fluidized bed contains n spherical particles of diameter d. If all the solute, or binder, which is introduced into the bed is distributed evenly such that each particle is coated with a layer of thickness a, then the mean granule diameter d' for a given mass of binder M_b is

$$d' = d + 2a \qquad\qquad 5.7$$

The volume of binder adhering to each core particle is equal to the difference in volume between spheres of diameter d' and d respectively, thus

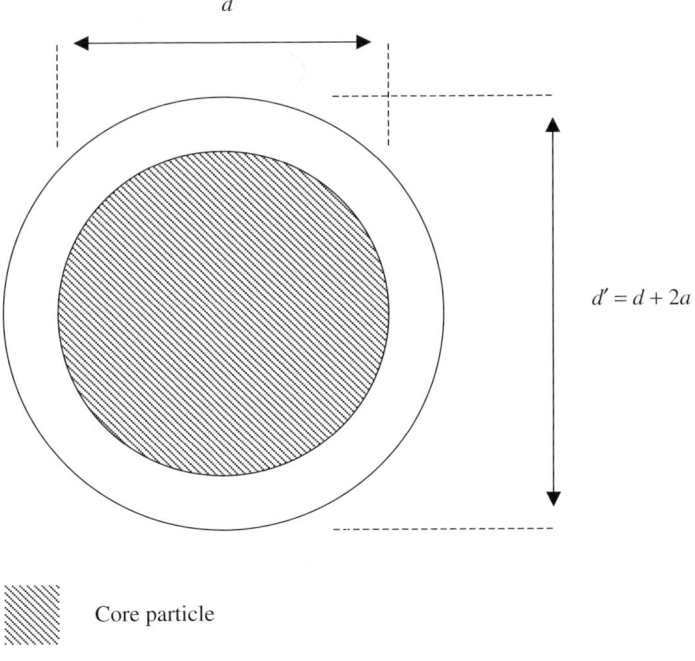

Core particle

Figure 5.7 Layered growth model.

$$\Delta V = V_{d'} - V_d \qquad 5.8$$

and

$$\Delta V = \frac{\pi}{6}(d + 2a)^3 - \frac{\pi}{6}d^3 \qquad 5.9$$

which becomes

$$\Delta V = \frac{\pi}{6}[8a^3 + 12a^2d + 6d^2a] \qquad 5.10$$

The mass m of a single core particle of density ρ_S is given by

$$m = \frac{\pi}{s}d^3\rho_S \qquad 5.11$$

and, if M is the total mass of bed particles, then

$$n = \frac{M}{m} \qquad 5.12$$

and the total number of particles in the bed is

$$n = \frac{6M}{\rho_S\pi d^3} \qquad 5.13$$

Now, if the volume of binder associated with each particle is V_b then

$$V_b = \frac{M}{\rho_b n} \qquad 5.14$$

Combining equations 5.13 and 5.14 gives

$$V_b = \frac{M_b\rho_S\pi d^3}{6M\rho_b} \qquad 5.15$$

and equating this with the expression for ΔV results in

$$4a^3 + 6a^2d + 3d^2a = \frac{M_b\rho_S d^3}{2M\rho_b} \qquad 5.16$$

Equation 5.16 can be solved for a, the layer thickness, from a knowledge of the total mass, density and mean diameter of the initial (core) bed particles. If values of M_b are known at different times then a, from equation 5.16, and d', from equation 5.7, can be found as a function of time.

Smith and Nienow (1983b) used this model to confirm the existence of particles resembling those represented by Figure 5.7. However, as these authors point out, the model is not intended to show that particles grow by the successive deposition of uniform layers of binder but rather that the final product approximates to a core particle surrounded by a layer of binder.

Agglomeration model

Originally proposed by Sherrington (1968) for moist agglomerates, this model has been adapted (Smith and Nienow, 1983b) to describe agglomerates of non-porous particles bound by solid bridges of material deposited from a drying solution (Figure 5.8). It is assumed that the fraction of void spaces between the initial particles which are filled with solid binder is f; in other words, this accounts for the possibility of entrapped air within the granule. Further, binder is withdrawn by a distance $s\,r$ (where r is the initial particle radius and s is an arbitrary parameter) into the interstices of the particles which lie at the granule surface. The particles from which the agglomerate is built are assumed to be spherical and of uniform radius. In an infinite well-packed mass of particles the void volume fraction is λ and the solid volume fraction is σ.

By definition

$$\lambda + \sigma = 1 \qquad\qquad 5.17$$

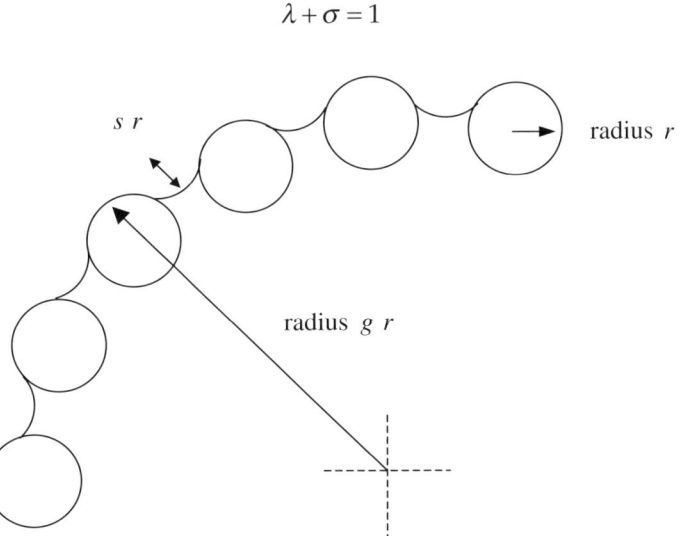

Figure 5.8   Agglomeration model: view of the granule surface.

Let the ratio of voids to solid be

$$k = \frac{\lambda}{\sigma} \qquad 5.18$$

The product granules are also assumed to be spherical and of uniform radius gr; the envelope volume V_g and external surface area S_g of a single granule are given by equations 5.19. and 5.20 respectively.

$$V_g = \frac{4\pi r^3 g^3}{3} \qquad 5.19$$

$$S_g = 4\pi r^2 g^2 \qquad 5.20$$

The ratio of total binder volume to total particle volume in the bed is denoted by y. If all granules are uniform this quantity must equal the binder volume to particle volume ratio for each granule. By definition, the volume of particles per granule is σV_g and the volume of binder per granule is $f\lambda V_g$ less the deficiency of binder at the granule surface, $\lambda S_g sr$. Therefore y is given by

$$y = \frac{f\lambda V_g - \lambda S_g sr}{\sigma V_g} \qquad 5.21$$

Substituting from equations 5.18 to 5.20 and rearranging yields

$$y = k\left(f - \frac{S_g sr}{V_g}\right) \qquad 5.22$$

or

$$y = k\left(f - \frac{3s}{g}\right) \qquad 5.23$$

Now, if β ($=1/g$), the ratio of initial particle diameter to granule diameter, is introduced then equation 5.23 becomes

$$y = k(f - 3s\beta) \qquad 5.24$$

This equation defines the relationship between the quantity of binder fed into the fluidized bed and the mean diameter of the product granules. A plot of y against β should give a straight line of gradient $-3\,k\,s$ and an intercept, at $\beta = 0$, equal to kf.

A theory of fluidized bed granulation

Particle growth mechanisms in fluidized bed granulation

Several authors (Jonke *et al.*, 1957; Hawthorn *et al.*, 1960; Markvart *et al.*, 1962) have suggested a particle growth mechanism for fluidized bed granulation, the essence of which is as follows. When the liquid feed is sprayed through an atomising nozzle into, or onto the surface of, a fluidized bed, discrete liquid droplets are formed which may either dry and form new discrete particles, or combine with existing bed particles in one of two ways. First, the liquid coats the particle surface, dries before a collision with a second particle is possible and consequently produces a growth layer of the dissolved feed substance. Second, wet particles coalesce and the liquid between them dries to form solid bridges and thus produces an agglomerate of two or more primary particles. In a continuous granulation system the equilibrium particle size will be determined by the balance between growth mechanisms, such as those outlined above, and mechanisms which lead to particle breakdown (Lee *et al.*, 1962; Ormos *et al.*, 1973a; Gonzalez and Otero, 1973) of which the most important are attrition and thermal shock (Buckham *et al.*, 1964; Bjorklund and Offutt, 1973; Nalimov *et al.*, 1975). Attrition is variously reported to be insignificant (Lee *et al.*, 1962), independent of major operating parameters such as gas velocity, feed concentration and atomising air rate (Gonzalez and Otero, 1973) or very significant and due largely to the effects of feed spraying (Buckham *et al.*, 1964) or of the fluidizing gas (Markvart *et al.*, 1962; Fukomoto *et al.*, 1970; Nikolaev *et al.*, 1975).

Gonzalez and Otero (1973), Lee *et al.* (1962) and Grimmett (1964) have all proposed a specific 'spray zone' of atomised liquid existing within a fluidized bed, through which the bed particles pass in a regular and ordered manner which are thus regularly and evenly coated with the feed liquid. This, it is suggested, gives rise to concentric growth rings around a core consisting of the original particle. This type of growth appears to be restricted to the work on high-temperature calcination and solution granulation. Pharmaceutical granulation, in which the object is to combine several different powders in the final granule, is concerned only with growth by agglomeration (with the exception of applying final coats to large drug particles) (Wurster, 1959; Wurster, 1960; Singiser *et al.*, 1966). Here the feed liquid is a binder solution prepared specifically to promote agglomeration (Ormos *et al.*, 1976).

It is far from certain that such a high-voidage spray zone can exist within a fluidized bed; a spray zone, with a submerged nozzle, would require a jet to be blown in the dense phase by the atomising air. Work by Rowe *et al.* (1979) and by Smith and Nienow (1982), using X-ray photography of fluidized beds, has shown that this does not occur. Plainly, atomisation of a liquid beneath the fluidized surface cannot be the same as atomisation into free air, and in the absence of a permanent high-voidage zone the often-quoted physical picture of small liquid droplets adhering to, and coating, single bed particles seems unrealistic. Bubbles, formed from either the fluidizing or atomising gas, may approximate to the required void zone but they will be periodic and the same arguments can be used against the theory when bubbles are not present. Further, should such a zone exist, the circulation of particles, although not entirely random because it is caused by relatively regular bubble motion, is far from ordered and the coating procedure which may take place in a spouted bed (see page 176) cannot occur. However, granules have been produced which consist of a core particle surrounded by deposited feed material. The subsequent models in the literature (see page 159) have described the product particles but have made no attempt to explain the precise mechanism by which such granules are produced.

Fluidizing gas velocity and particle mixing

Fluidizing gas velocity affects the extent of agglomeration and its magnitude is an important factor in determining whether or not a bed will quench. Particle circulation is proportional to the excess gas velocity and bed quenching is less likely if the solids circulation rate increases relative to the liquid feed rate. Better particle mixing results in improved liquid distribution and a reduced possibility of localised quenching. However, should quenching occur and substantial interparticle bonds or bridges form, higher gas velocities than for normal operation will be required to prevent segregation (Rowe and Nienow, 1976) and ensure that clumps of material do not build up at the bottom of the bed. It is also more probable, at higher velocities, that increased interparticle impacts, and impacts between particles and submerged surfaces, will result in greater abrasion and breakdown of agglomerates. Some slight improvement in heat and mass transfer between moist particles and the fluidizing gas can also be expected.

Binder properties

The nature of the feed liquid and its physical properties affects liquid distribution within the bed and thus the distribution of binder after solvent has evaporated. The viscosity of the feed liquid affects its atomisation characteristics and, for the same atomising air flow, more viscous liquids will give a larger droplet size (Masters, 1991). Solutions which become increasingly viscous as solvent evaporates may also have different distribution characteristics from those whose viscosity remains more or less constant. The distribution of binder is important in determining the type of granule produced; particles whose surfaces become entirely covered with liquid have a greater chance of drying before impacting with other particles and thus giving layered growth. Perhaps more probable and more importantly, their contacts with similarly coated particles are less likely to result in permanent bonds because less binder will be concentrated into the small area of contact.

The balance between granulation and fluidization

The successful operation of a fluidized bed granulator depends upon the balance between two essentially opposing factors. First, the binding mechanism which results in particles joining together to form larger ones because of the presence of liquid in the fluidized layer, and second, the 'disruptive' force – that is, the abrasive action of, and solids circulation within, the fluidized bed which tends to break down, or prevent the formation of, agglomerated particles. The magnitude and relative importance of these effects will depend upon, on the one hand, the quantity and physical properties of the liquid feed and, on the other, upon the characteristics of the fluidized bed such as the size and nature of the bed particles and the fluidizing gas velocity.

When a liquid, in any quantity, is introduced into a fluidized bed, liquid bonds are formed between individual bed particles, unless the particles are porous and capable of absorbing liquid. The formation of these bonds, which involves considerably stronger forces than either van der Waals or electrostatic effects, is inevitable because contact between two wet particles cannot be avoided in the dense phase. The extent and strength of these bonds will depend upon the amount of liquid available and its adhesiveness with the solid surface and the strength of the resultant solid bridges will be a function of the amount of deposited material and its intrinsic strength. In conventional granulators the mechanical action of the system helps in the binding process by kneading the materials together. In contrast, the particle motion in a fluidized bed acts against the binding mechanism and tends to control agglomeration and bond formation and consequently the particle size. The initial stages of agglomeration and bed quenching are identical and, for a given liquid feed, the fluidized bed parameters determine whether controlled particle growth takes place or whether the bed defluidizes. At one extreme it may be imagined that liquid sprayed into a packed bed, or a bed at the minimum fluidizing velocity, will result in a large, agglomerated mass of wet particles, whilst at the other, in a dilute-phase system, fewer particles will contact the liquid and those that do are much less likely to come together and form permanent bonds. Although these two elements do not act consecutively to produce a granular material (the physical picture is obviously far more complex), it is important to realise that the existence of the 'binding element' and the 'fluidization element' differentiates fluidized bed granulation from other rival granulation techniques.

Factors leading to bed quenching

The material and energy balances over a fluidized bed granulator must be satisfied if it is to operate successfully without wet quenching. Sufficient heat must be supplied to the bed, either through the bed walls or in the fluidizing gas, to provide the latent heat of vaporisation of the solvent, and the quantity of solvent evaporated must not exceed that which will saturate the off-gases at the operating temperature. Failure to meet either of these requirements results in excess liquid in the fluidized layer and therefore wet quenching. Clearly there is a limit to the amount of liquid that can be tolerated in the bed and beyond which operation becomes impossible. This excess need not be over the whole bed but may occur in a localised region, for example close to the nozzle or 'feed zone'. Localised wet quenching in this manner results in large clumps or agglomerates which then segregate at the bottom of the bed. Once this has happened, and the bed is partly defluidized, loss of important fluidized bed characteristics (such as particle mixing and good heat transfer) quickly follow, leading to further agglomeration and complete failure of the process. This is certainly the case if the bed is not sufficiently well fluidized to break up agglomerates as they form. Whether wet or dry quenching occurs depends on the rate of drying of the feed liquid and therefore its concentration and the bed temperature.

Other than gas velocity and the physical properties of the feed, particle size is a parameter which has a significant effect. Smaller bed particles are more likely to form permanent bonds, and to quench, because of their smaller inertia. The force tending to pull two particles apart is equal to the product of the particle mass and the distance between the two centres of mass. For the case of two spherical particles joined together at their surfaces, this force is proportional to the particle diameter raised to the fourth power. Other cases approximate to this relationship.

An overall mechanism

As outlined above, the initial stages of agglomeration and of bed quenching are exactly the same. However, Smith and Nienow (1983a) proposed that all modes of growth and bed quenching have the same initial stage, i.e. the formulation of liquid bonds between adjacent particles in the fluidized bed. When solvent is evaporated from the feed solution liquid bonds give rise to solid bridges between those same adjacent particles unless there is a redistribution of binder either before or after the solution dries. Whether this redistribution takes place by a breaking of either liquid bonds or solid bridges depends on the balance between the two elements of fluidized bed granulation outlined above.

The distribution of the feed liquid and of binder throughout the bed and on the surface of individual particles depends upon the structure of the bed particles and the viscosity of the feed solution. A binder solution which increases rapidly in viscosity as solution is evaporated, and the concentration of binder increases, is less likely to be able to flow around a bed particle and cover the surface area and consequently layered particle growth is much less probable. A solution for which the viscosity remains more or less constant with concentration is better able to spread around a particle before the solution dries and forms a solid crust. However, the intensity of particle-particle contacts makes it extremely unlikely that significant coverage of the surface will occur.

The existence of intraparticle porosity in the bed allows liquid to be evaporated over a larger proportion of the bed than is possible when non-porous particles are used. Consequently temperature gradients are far less pronounced and there is a reduced possibility of wet quenching with generally more stable operation. The viscosity of the feed solution is an important factor in determining the behaviour of porous bed particles. A higher viscosity solution is less able to flow into intraparticle pores and the no-growth period is considerably shorter because, with an increased viscosity, the time taken for solution to enter the pores is much greater and it dries before significant penetration is achieved. The pores become blocked more quickly, with a smaller fraction of the pore volume filled, and liquid bonds begin to form on the exterior particle surface far earlier than is the case with a lower viscosity solution. Once the pores are blocked, porous and non-porous particles behave in a similar manner.

When the fluidized bed consists of non-porous particles, solvent in the feed liquid is evaporated in a well-defined zone close to the spray nozzle and from the surface of the bed particles with which it inevitably comes into contact. No permanent gas jet or void exists in this region; particle motion is not well ordered and no regular coating of particles with feed solution takes place. The random and intense contact between particles and liquid results in agglomeration. Even if the mass and heat balances for the bed as a whole have been satisfied, both localised wet or dry quenching may occur in this zone. In the latter case, as the agglomerated mass which is produced begins to move away from the zone where liquid first contacts the bed particles, drying of the solution takes place. If the dry mass is not broken down and reduced to smaller packets of agglomerated particles, it segregates as an effectively very large particle and sinks to the bottom of the bed.

Smith and Nienow (1983a) observed dry rather than wet quenching, evidenced by a rise in temperature at the bottom of the bed when a segregated layer was formed. The deposition of larger amounts of solute from the same volume of feed solution results in a bridge between two particles of greater strength, bond strength being proportional to the product of intrinsic binder strength and the quantity of binder present. Thus, a high mass flow rate of binder (and therefore a high solution concentration) will dominate the fluidization/granulation balance and intensify the quenching problem, as the aggregated mass is less able to break down.

The extent to which clumps of agglomerated particles remain intact determines the outcome of the fluidized bed granulation process; it governs the type and size of the granular material which is produced. Bed quenching results if insufficient breakdown takes place; and breakdown into smaller agglomerates, to an equilibrium size, results in agglomerated granules. Further reduction and tearing apart of smaller agglomerates ultimately produces a single bed particle with associated binder; in other words, a layered or coated granule.

For a given fluidizing gas velocity and a given initial bed particle size, i.e. a constant fluidization element, the relative dominance of the granulation element over fluidization will be a function of bond strength, and of intrinsic bond strength if the binder concentration is constant. The stronger the solid matter between particles (either pairs of particles or particles in a large mass), the less is the breakdown of that agglomerate. Weak bonds are torn apart more easily, by fluid drag on the bed particles and by abrasion in the fluidized layer, leaving the binder behind attached to the surface of one or both of the particles which it bound together. As agglomerates break down to single particles, the attached binder causes an increase in size of the initial particles. Constant repetition of this process, the reagglomeration of particles with fresh feed liquid followed by breakdown (before or after complete drying has taken place), produces a growth layer. Smith and Nienow (1983a) argued that the absence of concentric growth rings, or of spherical product granules, strongly suggests that growth does not occur by a mechanism of regular and uniform coating with successive layers of feed material. The appearance of the layered granules in their work was consistent with the random deposition of binder which would result from continual formation, breakage and re-formation of liquid bonds and solid bridges, on the particle surface. Subsequent redissolution of binder and particle attrition produces a less angular product.

For a particular binder, the mode of particle growth, and the growth rate, depend upon the fluidization variables: gas velocity and particle size. The fluidizing gas velocity contributes to two effects. First, particle circulation increases at higher excess gas velocity and as it is increased in relation to the solution feed rate, the amount of feed associated with each particle decreases. Bed quenching therefore becomes less probable, allowing higher feed rates than at lower gas velocities. Second, once interparticle bonds or bridges have formed they break down more easily at higher velocities because of increased fluid drag and increased abrasion of granules due to the greater number of particle-particle impacts occurring. When larger initial particles were used, Smith and Nienow (1983a) showed that either the mean diameter of granules decreased or layered growth occurred under conditions which otherwise lead to quenching.

Food applications of fluidized bed granulation

Instantising

A major application of fluidized bed granulation is in the production of instant products in which primary particles are agglomerated to give a granule with improved wettability, dispersion and dissolution compared to the original primary particles. Such materials, typically milk powders (Kang and Shin, 2004), are produced initially from a liquid feed by spray drying followed by treatment in a fluidized bed which is integrated in the production line. Masters (1983) classifies such processes as either (1) a straight-through process or (2) a re-wetting process. In the former, the spray drier is operated at a lower outlet temperature than normal so that the powder leaving the spray drying chamber has higher moisture content. This then enters a vibrating fluidized bed where self-adhesion, or self-agglomeration, of the particles occurs because of the presence of surface moisture and final drying of the granules takes place. A second fluidized bed is used to cool the product (Figure 5.9) which is characterised by coarse granules which are free of fines. A straight-through instantiser typically is used for the production of instant skim milk although other uses include the production of non-caking whey powder, protein powders, instant beverage whiteners and baby food formulae (Masters, 1991).

The re-wetting process (Figure 5.10) gives a higher degree of agglomeration and better instant properties (Masters, 1991). A number of variants on the basic idea exist (Masters, 1983).

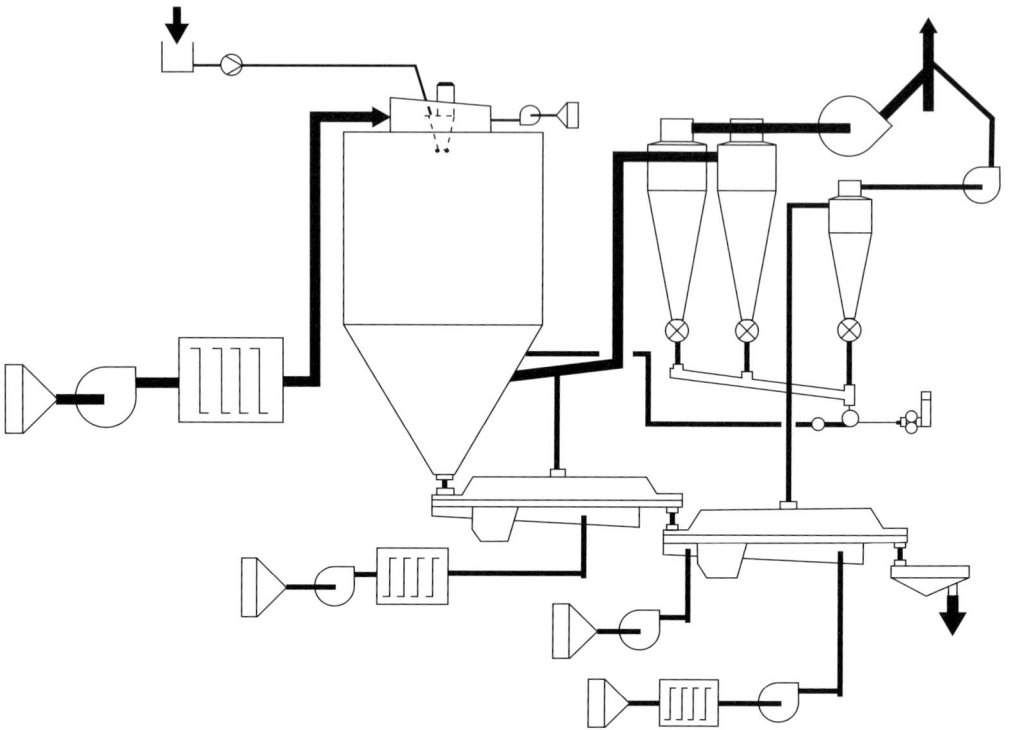

Figure 5.9 Straight-through instantising process. Reprinted from Masters (1983) with permission.

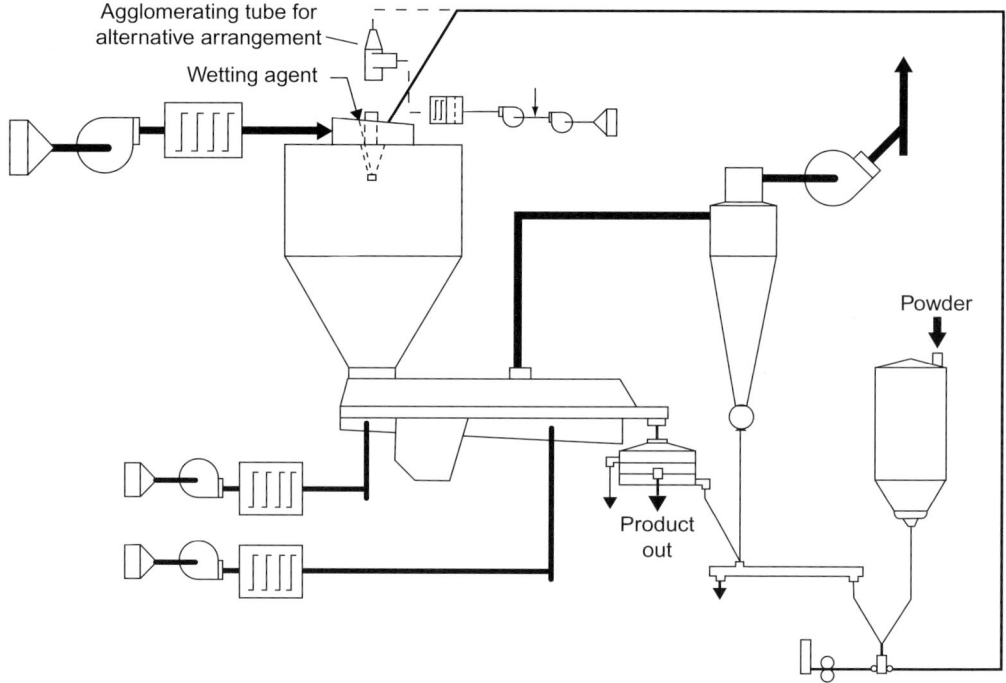

Figure 5.10 Re-wetting instantising process. Reprinted from Masters (1983) with permission.

(1) Powder is allowed to fall through a vertical tube positioned at the top of a large chamber, and into which steam is injected. Cool air is drawn into the tube and thus condenses the steam onto the powder surface. The resultant agglomerates are 'stabilised' within the chamber before falling into a vibrated fluidized bed for final drying and cooling. This is referred to as surface agglomeration.

(2) In so-called droplet agglomeration, atomised water is sprayed onto the powder using a rotary (or spinning disc) atomiser as it passes through the vertical tube. Alternatively, twin-fluid nozzles may be used if the tube is mounted at the inlet to the fluidized bed rather than at the chamber inlet. In either case the water may be replaced with lecithin or with other binding agents such as solutions of flavour compounds. This second alternative (the addition of a wetting agent at the fluidized bed inlet) is used when lecithin is employed to coat whole milk powder in the production of instantised whole milk. Perez and Westergaard (1995) describe such a process in detail.

(3) In a process known as rotating disc agglomeration, powder from the spray drier is wetted as in the above methods but then falls onto a rotating disc and agglomerates whilst in contact with the disc. Although subsequent drying takes place in a vibrated fluidized bed, this cannot be classified as true fluidized bed granulation. Such a process is used for instant breakfast mixes, chocolate beverages and instant coffee.

Instantisation, in which the fluidized bed is responsible for a significant change to the structure and size of the granules, should not be confused with the use of an integrated fluidized bed drier immediately following spray drying. There are many descriptions in the literature of such processes; a recent example of a novel application of this technique (Jha *et al.*, 2002) is in the production of a shelf-stable powdered kheer mix, an Indian dessert which consists of a sweetened mixture of cooked partially concentrated milk and rice flour. This is spray dried, the particles are then dried in a fluidized bed and finally dry blended with sugar.

Encapsulation and coating

The coating of food particulates, often referred to as encapsulation, has a wide range of purposes including increasing shelf-life, masking taste or odour, improving appearance and colour and improving ease of handling. Layered or coated granules can be designed to control the release of constituents within the particle (Dewettinck and Huyghebaert, 1998) by providing a coating which dissolves at a given rate in a particular environment, for example at a particular temperature or pH, or to protect unstable ingredients from degradation by heat, moisture or light (Teunou and Poncelet, 2002). The coating materials applied for food applications are mainly water-soluble biopolymers and include lipids and edible films based on either milk proteins or corn proteins (Dewettinck *et al.*, 1998). Dewettinck and co-workers report experimental work using gums, such as locust bean gum, carboxymethylcellulose (CMC), sodium alginate or kappa-carrageenan (Dewettinck *et al.*, 1998), gelatin and starch hydrolysate (Dewettinck *et al.*, 1999a) and protein concentrates such as sodium caseinate, lysozyme and blood plasma concentrate (Dewettinck and Huyghebaert, 1998).

Lipid coating (sometimes referred to as hot melt coating) of food particles is used for colouring, surface protection, taste and odour masking and the controlled release of vitamins, salts, preservatives and food additives (Eichler, 1996). Eichler (1996) reports that top, bottom and tangential spraying have all been used but that top spraying, presumably onto the bed surface, is best suited to hot melt coating. Dewettinck *et al.* (1999b) suggest that top spraying is most appropriate for food applications largely because of its versatility (both coating and agglomeration are possible) and simplicity. The lipid melting point (which should be between 40°C and 80°C), the melting range and the viscosity of the feed liquid are crucial (Eichler, 1996). Optimal spreading of the molten droplets requires the coating to congeal slowly and therefore the bed temperature must be as close as possible to the melting point of the lipid coating (and certainly within 10°C). However, if the bed temperature becomes too close to the melting point the product will become too sticky and agglomeration and quenching will result. Eichler (1996) further suggests that the lipid should be heated to between 40°C and 60°C above its melting point before being sprayed and therefore the atomisation air must be heated to prevent solidification. Lipid coatings with melting points below 40°C are not recommended because of the possibility of exposure to ambient temperatures approaching this level during transport or handling of the product.

Dewettinck *et al.* (1999b) suggest that a bed used for coating should be operated at the 'thermodynamic operation point' which they define in terms of the steady-state outlet bed air temperature and the outlet relative humidity. These authors present a model for optimisation of this condition which they tested experimentally by spraying distilled water on to 365 μm diameter glass beads. Good agreement between model and experiment is reported when heat losses from the bed are included in the calculation.

Teunou and Poncelet (2002), in their extensive review of fluidized bed coating, identify three geometries for continuous processing. First, the monocellular bed which is essentially the type of equipment used

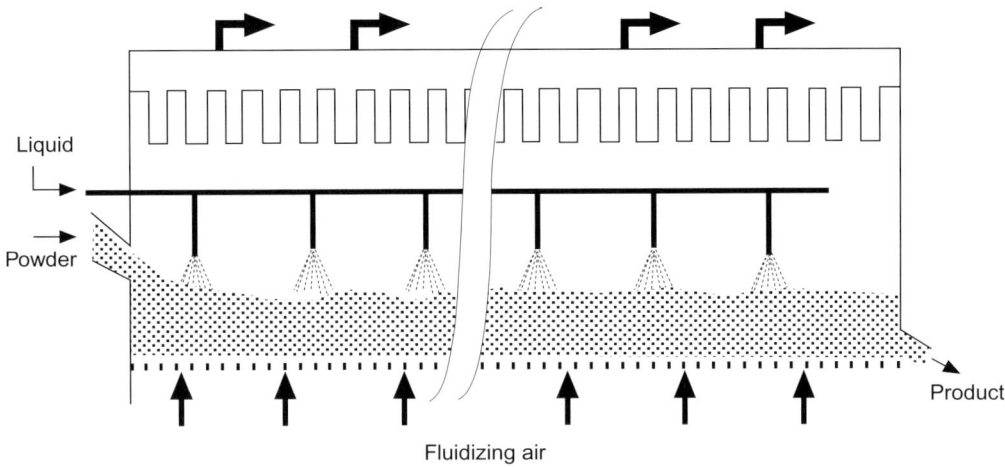

Figure 5.11 Real horizontal fluidized bed granulator. Reprinted from Teunou, E. and Poncelet, D., Batch and continuous fluid bed coating: review and state of the art, *J. Food Eng.*, **53** (2002) 325–340, with permission from Elsevier.

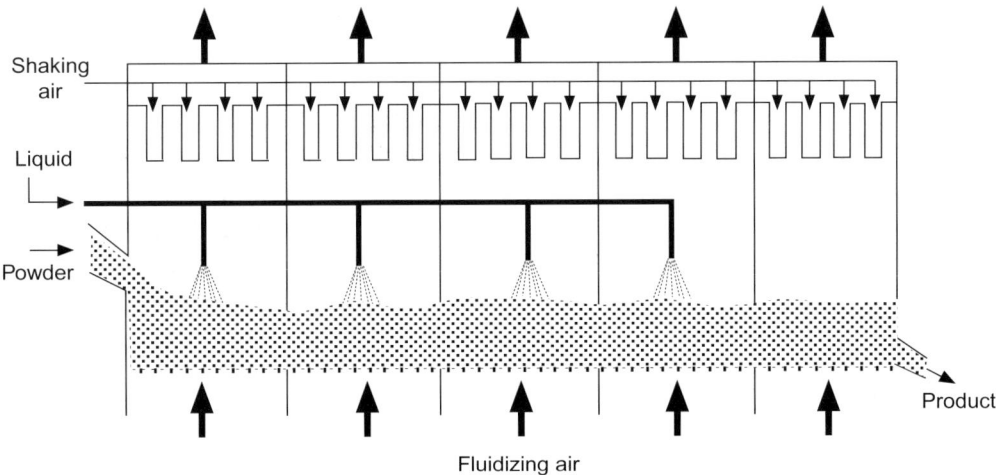

Figure 5.12 Multicell fluidized bed granulator. Reprinted from Teunou, E. and Poncelet, D., Batch and continuous fluid bed coating: review and state of the art, *J. Food Eng.*, **53** (2002) 325–340, with permission from Elsevier.

for batch processing (i.e. usually a circular cross-section bed) but with an inlet for the powdered feed and a product outlet. Second, what these authors term a 'real' horizontal fluidized bed. This is a single compartment of rectangular cross-section (Figure 5.11), with no obstacles or weirs in the bed, through which material passes continuously, often with the aid of vibration of the entire bed (Kage *et al.*, 1996). Finally, the multicell bed is divided into four or five compartments along its length with different fluidizing conditions in each (Figure 5.12).

A wide range of foods, food ingredients and food additives has been coated or encapsulated using fluidized bed techniques, including enzymes, vegetable proteins, yeast and bacteria (Teunou and Poncelet, 2002); confectionery and nuts (Casimir *et al.*, 1968); peanut pieces (Kang and Shin, 2002) and dietary fibre coated with a hemicellulose derived from grain hulls (Antenucci, 1990). Rümpler and Jacob (1998) applied a molten fat coating to a very hygroscopic crystalline confectionery product, in a continuous operation, to give a free-flowing product at room temperature which improved product storage and prevented agglomeration.

Other applications

A large number of the reported applications of fluidized bed granulation in the food industry relate to the production of beverages, including beer. Many of these processes are the subject of patents. Peterreins and Kamil (2002a) patented a fluidized bed process (the 'PlatoTec vacuum evaporation/fluidized bed spray granulation process') for the manufacture of a wort granulate, used in the brewing of beer, which can subsequently be re-dissolved and fermented conventionally. A further patent (Peterreins and Kamil, 2002b) describes the addition of one or more layers of coating materials, especially flavourings, to wort granulates. These added flavourings may be encapsulated to avoid losses by volatilisation during storage and transport. Piotrowski and Kamil (2002) suggest further advantages: that the process can be used for the manufacture of alcohol-free beer, and that a wide range of flavour and colour variants may be manufactured. The process permits the production of wort concentrate at a central location followed by dissolution of the wort granules and continuous conventional processing at localised breweries (Kamil and Peterreins, 2001); the wort granulate is reported to be stable, easy to transport, dust free, and easy to dissolve. The use of dry wort granules in this way allows both the brewing of identical beers at a number of locations and the production of a range of beers in an individual brewery.

A process for the production of granular cocoa in a fluidized bed has been described (Kimura and Terauchi, 1998, 1999) in which the liquid binder is prepared by heating cocoa powder and/or cocoa extract in water. The resultant granules have a moisture content of 15–40% by weight and are then dried in the same fluidized bed. The granular cocoa is reported to dissolve easily in warm milk or warm water without preliminary mixing. Kirchmann (1996) reports a similar process for the agglomeration of mixtures of cocoa powder, dried milk and sugar for use in vending machines. Camp and Fischbach (1990) agglomerated a powdered beverage base consisting of cocoa, coffee,

skim milk powder and sugar followed by coating with molten chocolate. The fluidized bed was maintained at a temperature of 20–60°C
during coating followed by cooling to harden the chocolate. A further
example of a coating process is the manufacture of candied roasted
coffee beans (Winkelmann *et al.*, 2000) in which the beans are roasted
in a fluidized bed for between 70 and 200 s at 200–270°C using a mixture
of steam and other gases. After roasting, crystalline sugar or sugar
solution (at a concentration of between 2 and 4 kg sugar per kg water)
is dispersed into the fluidized bed. The sugar dries and caramelises on
the surface of the coffee beans to form a uniform layer over each bean.
The coffee beans are then cooled in the fluidized bed using ambient
air over a period of 110–400 s. Other applications of fluidized bed
granulation include the granulation of tea using a tea extract preparation as binder (Guiua *et al.*, 1990).

Kovacova (1990) describes a method for the batch drying of fruit and
vegetable pulp in a fluidized bed in which carrier particles (variously
crystalline and caster sugar, dried skim milk, potato and wheat starch,
apple powder, semolina or oat flakes), pre-moistened to a solids content
of between 55% and 76%, are fed to a preheated fluidized bed and
sprayed with the pulp to be dried. A product with a narrow particle
size distribution and a uniform pulp content is claimed. Specific foods
for which fluidized bed granulation has been used include potato
puree (Zelenskaya and Pilipenko, 1989) and granulated dried apple
(Haida *et al.*, 1994).

Caspers *et al.* (2001) proposed the use of a multi-zone fluidized bed
for the production of a gelling aid consisting of particulate sugar, finely
comminuted particles of pectin and a granular acidulant such as citric
acid. In the first zone of the bed the fluidized sugar and pectin are
agglomerated by top spraying with water. In the second zone a liquid
coating is applied which is dried in the third zone. Finally, the granulated product is mixed with granular citric acid. The gelling aid is
claimed to be resistant to clumping and abrasion, stable against demix-
ing and has a long shelf-life. Berizzi (2004) patented a process in which
the protein-rich fraction of flour is granulated with water and added
to baking flour to improve its processing properties and the quality of
bakery products.

Lutz and Paprotny (1989) granulated a curd, prepared by rennet
coagulation of milk, adding flavourings, seasonings, food colours,
sweeteners or preservatives either before, during or after the granulation process. The granules are used as structure-forming agents or as
carbohydrate or fat substitutes, in chocolate, confectionery, bakery
products and beverage mixes. A process in which it is claimed that
particles are built up layer by layer has been proposed for the granulation of varying compositions of citric, malic and fumaric acids, thus

allowing for variation in the tartness of flavour profiles of such additives (Anon., 1995). Buffo *et al.* (2002) studied the effects of agglomeration on physicochemical properties of encapsulated spray-dried flavourings. Model flavouring systems consisting of orange oil and ethyl esters were encapsulated by spray drying in traditional carrier materials (modified starch, maltodextrins or gum acacia) with subsequent agglomeration in a fluidized bed. Compared with encapsulated controls (no agglomeration), agglomerated flavourings had decreased surface oil and higher absolute density, mean particle size and moisture contents. Agglomeration resulted in a slight flavour loss.

Feed additives produced during fermentation in the liquid phase are conventionally separated from fermentation broth during further downstream processing. Walter and Rümpler (2000) described the development of a fluidized bed granulator to convert the entire fermentation liquid into a free-flowing, dust-free, granular powder with a final moisture content of 2% and a mean particle size below 1 mm, although milling and recycle of oversized particles was necessary to meet this specification. The fermentation liquid contained between 20% and 30% solids and this was increased to 50% before granulation in a 2.2 m diameter production unit with three spray nozzles placed in the bottom of the bed. The bed mass was 700 kg and fermentation liquid was fed to the unit at a spray rate of 880 kg h^{-1}. The authors report that fluidized beds up to 3 m diameter and corresponding water evaporation rates of 5000 kg h^{-1} are possible.

The starch processing, dairy, brewing and baking industries are users of bulk quantities of a wide range of enzymes which are usually sold as concentrated liquids or as solid powders with bulking agents such as sodium chloride. A major problem associated with dry enzyme preparations is their high dust level and consequent problems of allergic reactions. Härkönen *et al.* (1993) granulated a range of fine enzyme powders in an Aeromatic fluidized bed granulator at the 1 kg batch scale, using sodium chloride as a filler and modified potato or tapioca starch hydrolysate as a binder. The mean particle size of the granules ranged from 135 to 200 μm. Other applications have included the production of artificial sweeteners (Fotos and Bishay, 2001), a granular flavouring for chewing gum which results in enhanced flavour retention (Hyodo *et al.*, 2003) and a free-flowing granular dried soup mix with a relatively narrow particle size distribution to aid dispersion (Haefliger *et al.*, 2001).

Spouted bed granulation

Granulation and drying of solutions in a spouted bed were first demonstrated by Berquin (1961). Particle motion which is regular and ordered (far more so than in a fluidized bed; see Chapter 1), large particles and high gas velocities are all characteristic of a spouted bed. Solution may be sprayed into the bed at the gas inlet or onto the bed surface. Mathur and Epstein (1974) point out the advantages of this type of equipment: it is suitable for continuous operation, gives a product of near uniform size and allows particles to grow much larger than in a fluidized bed granulator. The high-voidage, high-temperature zone near the gas inlet allows very rapid evaporation of solvent and, together with the cyclic particle motion and the effect of large bed particles, results in very low rates of agglomeration.

Two early papers identified the mechanisms of particle growth. Uemaki and Mathur (1976) granulated ammonium sulphate and recorded growth over periods of up to nine hours, with granules between 1 and 4 mm in diameter being produced. In similar apparatus Robinson and Waldie (1979) produced sodium chloride granules up to 7 mm in diameter from a 23% feed solution. Both sets of workers concluded that the dominant mechanism was growth by continuous deposition and layering of solute on the seed particles, although the evidence presented for the relative significance of competing mechanisms, and for size dependency of growth, is contradictory. An alternative technique is to spray the feed solution onto a bed of inert particles. The solute dries as a layer around the particle and is then removed by attrition. The dried matter becomes entrained in the exhaust air and is separated in a cyclone. Tia *et al.* (1995) used this method for the drying of rice flour slurry, a coating of which was formed initially on 5 mm diameter ceramic spheres, whilst Shuhama *et al.* (2003) dried anatto powder from solution using 2.6 mm diameter glass spheres. A variation on this was proposed by Szentmarjay and Pallai (1989). Their apparatus consisted of a spouted bed with a vertical screw conveyor positioned along the axis of the bed. The screw diameter was approximately equal to that of the air inlet at the base of the cone. This arrangement allows the air flow rate to be selected on the basis of drying requirements and not on the need to maintain particle motion. Further, the variation in screw speed allows some control of particle circulation time and therefore control of the abrasion of dried material. Szentmarjay *et al.* (1996) used the technique to dry baker's yeast and tomato concentrate.

Nomenclature

a	thickness of growth layer
c	heat capacity at constant pressure
d	diameter of constituent particles
d'	layered granule diameter

f	fraction of the granule void volume filled with binder
g	ratio of granule diameter to initial particle diameter
h_{fg}	latent heat of vaporisation
H_i	absolute humidity of fluidizing gas
H_o	absolute humidity of exhaust gas
k	ratio of voids to solid
m	particle mass
M	bed mass
M_b	mass of binder
n	number of fluidized bed particles
Q_W	heat supplied through bed wall
Q_L	rate of heat loss
r	initial particle radius
s	arbitrary parameter
S_g	granule external surface area
T	temperature
T_i	inlet gas temperature
T_o	exhaust gas temperature
u	superficial gas velocity
u_{mf}	minimum fluidizing velocity
V	particle volume
V_b	volume of binder per particle
V_g	granule envelope volume
W_A	mass flow rate of fluidizing gas
w	mass flow rate of feed liquid
x_s	mass fraction of solute in feed liquid
y	ratio of total binder volume to total particle volume in the bed

Greek symbols

β	ratio of initial particle diameter to granule diameter
σ	tensile strength of a capillary state granule; tensile strength of an agglomerate; solid volume fraction of agglomerate
γ	surface tension
ε	interparticle voidage within a granule
θ	intrinsic tensile strength of solid bridge
λ	void volume fraction of an agglomerate
ρ_S	core particle density
ρ_b	binder density

Subscripts

A	fluidizing gas
V	vapour

F feed

S solute; solid product stream

References

Anon., Citric acid is no lemon, *Food Review*, **22** (1995) 51, 53.

Antenucci, R.N., Method of making a hemicellulose coated dietary fiber, United States Patent, US4927649, 1990.

Bakhshi, N.N. and Chai, C.Y., Fluidized bed drying of sodium sulphate solutions, *Ind. Eng. Chem. (Proc. Des. Dev.)*, **8** (1969) 275–279.

Bakhshi, N.N. and Nihilani, A., Fluidized bed drying of sodium sulphate solutions; fluidisation and its applications. Proceedings of an International Symposium on Fluidisation, Toulouse, 1973, 534–544.

Barsukov, E.Y. and Soskind, D.M., Particle grinding in a fluidized bed, *Int. Chem. Eng.*, **13** (1973) 84–86.

Berizzi, H.P., Process for manufacture of flour mixtures for baking, German Federal Republic Patent DE10248160A1, 2004.

Berquin, Y.F., A new granulating process, *Gen. Chim.*, **86** (1961) 45.

Bjorklund, W.J. and Offutt, G.F., Fluidized bed denitration of uranyl nitrate, *A.I.Chem.E. Symp. Ser.*, **69** (128) (1973) 123–129.

Buckham, J.A., Lakey, L.T. and McBride, J.A., Calcination of aluminium nitrate wastes, *Chem. Eng. Prog. Symp. Ser.*, **60** (53) (1964) 20–31.

Buckham, J.A., Ayers, A.L. and McBride, J.A., Fluidized bed calcination of high level radioactive waste in a plant scale facility, *Chem. Eng. Prog. Symp. Ser.*, **62** (65) (1966) 52–63.

Buffo, R.A., Probst, K., Zehentbauer, G., Luo, Z. and Reineccius, G.A., Effects of agglomeration on the properties of spray-dried encapsulated flavours, *Flavour and Fragrance J.*, **17** (2002) 292–299.

Camp, W.F. and Fischbach, E.R., Chocolate coated beverage mixes, United States Patent, US4980181, 1990.

Capes, C.E., Particle size enlargement, Elsevier, London, 1979.

Capes, C.E. and Danckwerts, P.V., Granule formation by the agglomeration of damp powders, *Trans. Inst. Chem. Engrs.*, **43** (1965) 116–123.

Casimir, D.J., McBean, D. and Shipton, J., Fluidization techniques in food processing, *Food Tech. Australia*, **20** (1968) 466–469.

Caspers, G., Klein, C., Krell, L. and Moerle-Heynisch, T., Process for manufacture of a gelling sugar, German Federal Republic Patent, DE10016906A1, 2001.

Crooks, M.J. and Schade, H.W., Fluidized bed granulation of a microdose pharmaceutical powder, *Powder Tech.*, **19** (1978) 103–108.

D'Amore, M., Donsi, G. and Massimilla, L., The influence of bed moisture on fluidization characteristics of fine powders, *Powder Tech.*, **23** (1979) 253–259.

Davies, W.L. and Gloor, W.T., Batch production of pharmaceutical granulations in a fluidized bed, I. Effects of process variables on physical properties of final granulation, *J. Pharm. Sci.*, **60** (1971) 1869–1874.

Davies, W.L. and Gloor, W.T., Batch production of pharmaceutical granulations in a fluidized bed, II. Effects of various binders and their concentrations on granulations and compressed tablets, *J. Pharm. Sci.*, **61** (1972) 618–622.

Davies, W.L. and Gloor, W.T., Batch production of pharmaceutical granulations in a fluidized bed, III. Binder dilution effects on granulation, *J. Pharm. Sci.*, **62** (1973) 170–171.

Dewettinck, K. and Huyghebaert, A., Top-spray fluidized bed coating: effect of process variables on coating efficiency, *Lebensmittel Wissenschaft Technologie*, **31** (1998) 568–575.

Dewettinck, K., Deroo, L., Messens, W. and Huyghebaert, A., Agglomeration tendency during top-spray fluidized bed coating with gums, *Lebensmittel Wissenschaft Technologie*, **31** (1998) 576–584.

Dewettinck, K., Deroo, L., Messens, W. and Huyghebaert, A., Agglomeration tendency during top-spray fluidized bed coating with gelatin and starch hydrolysate, *Lebensmittel Wissenschaft Technologie*, **32** (1999a) 102–106.

Dewettinck, K., De Visscher, A., Deroo, L. and Huyghebaert, A., Modeling the steady-state thermodynamic operation point of top-spray fluidized bed processing, *J. Food Eng.*, **39** (1999b) 131–143.

Eichler, K., Lipid coating in the food industries, *Food Tech. Europe*, **3** (1996) 122–126.

Fotos, J. and Bishay, I., Process for preparing an N-[N-(3,3-dimethylbutyl)-L-alpha-aspartyl]-L-phenylalanine 1-methyl ester agglomerate, United States Patent, US6180157B1, 2001.

Frantz, J.F., The evaporation of brine solutions in a fluidized salt bed, PhD thesis, Louisiana State University, 1958.

Fukomoto T., Maeda, K. and Matagi, Y., Studies on the fluidized bed calcination of radioactive liquid wastes, *Nucl. Sci. Eng.*, **7** (1970) 137–144.

Gluckman, M.J., Yerushalmi, J. and Squires, A.M., Defluidisation characteristics of sticky or agglomerating beds, in: Keairns, D.L., (ed.), Fluidization technology volume 2. Proceedings of an International Conference on Fluidization, Pacific Grove, California, 1975. Hemisphere, New York, 1976, 395–422.

Gonzalez, V. and Otero, A.R., Formation of UO_3 particles in a fluidized bed, *Powder Tech.*, **7** (1973) 137–143.

Grimmett, E.S., Kinetics of particle growth in the fluidized bed calcination process, *A.I.Chem.E.J.*, **10** (1964) 717–722.

Guiua, K.P., Revishvili, T.O., Maisuradze, Z.A. *et al.*, Production of granulated tea, USSR Patent, SU1595428, 1990.

Haefliger, H., Dupart, P., Blasius, L. and Kehrli, P., Granular food product, European Patent, EP1074188A1, 2001.

Haida, H., Kroyer, G.T., Kuenne, H.J., Washuettl, J. and Winker, N., Anwendung der Wirbelschichttechnologie zur Herstellung eines Apfeltrockenproduktes [Use of fluidized bed drying for manufacture of a granulated dried apple product], *Deutsche Lebensmittel Rundschau*, **90** (1994) 9–15.

Härkönen, H., Koskinen, M., Linko, P., Siika-aho, M. and Poutanen, K., Granulation of enzyme powders in a fluidized bed spray granulator, *Lebensmittel Wissenschaft Technologie*, **26** (1993) 235–241.

Hawthorn, E., Shortis, L.P. and Lloyd, J.E., Fluidised solids drying process for production of uranium tetrafluoride, *Trans. Inst. Chem. Engrs.*, **38** (1960) 197–215.

Hyodo, M., Fujimoto, K., Abe, S. and Tokizane, M., Chewing gum composition, United States Patent, US6537595B1, 2003.

Jha, A., Patel, A.A. and Singh, R.R.B., Physico-chemical properties of instant kheer mix, *Lait*, **82** (2002) 501–513.

Jonke, A.A., Petkus, E.J., Loeding, J.W. and Lawroski, S., Calcination of dissolved nitrate salts, *Nucl. Sci. Eng.*, **2** (1957) 303–319.

Kage, H., Oba, M., Ishimatsu, H., Ogura, H. and Matsuno, Y., The effects of frequency and amplitude on the powder coating of fluidized particles in vibro-fluidized bed, *J. Soc. Powder Tech. Japan*, **33** (1996) 711–716.

Kamil, G. and Peterreins, F., Separation of wort production from the rest of the brewing process using fluidized bed technology, Brauwelt, **141** (2001) 1338, 1340–1341.

Kang, H.A. and Shin, M.G., Effect of inlet air temperature and atomizing air pressure on fluidized bed coating efficiency of broken peanut, *Korean J. Food Sci. Tech.*, **34** (2002) 924–926.

Kang, H.A. and Shin, M.G., Optimization of fluidized bed granulating conditions for powdered milk by response surface methodology, *J. Korean Soc. Food Sci. Nut.*, **33** (2004) 225–228.

Kapur, P.C. and Fuerstenau, D.W., Size distributions and kinetic relationships in nuclei region of wet pelletization, *Ind. Eng. Chem. (Proc. Des. Dev.)*, **5** (1966) 5–10.

Kimura, Y. and Terauchi, M., Process for producing granular cocoa. European Patent, EP0885567A2, 1998.

Kimura, Y. and Terauchi, M., Process for producing granular cocoa, United States Patent, US6007857, 1999.

Kirchmann, A., Agglomeration von Kakaomischungen [Agglomeration of cocoa mixtures], *Zucker und Süßwarenwirtschaft*, **49** (1996) 139–140.

Kovacova, S., Modern technique for drying fruits and vegetables, *Prumysl Potravin*, **41** (1990) 539–541.

Kurita, Y. and Sekiguchi, I., Effect of vortex orifice air distributor on granule growth in conical fluidized bed granulation with bottom entry spray, *J. Chem. Eng. Japan*, **33** (2000) 57–66.

Lee, B.S., Chu, J.C., Jonke, A.A. and Lawroski, S., Kinetics of particle growth in a fluidized bed calciner, *A. I. Chem. E. J.*, **8** (1962) 53–58.

Legler, B.M., Feed injector for heated fluidised beds, *Chem. Eng. Prog.*, **63** (2) (1967) 75–82.

Linkson, P.B., Glastonbury, J.R. and Duffy, G.J., The mechanism of granule growth in wet pelletising, *Trans. Inst. Chem. Engrs.*, **51** (1973) 251–259.

Lutz, C. and Paprotny, S., Verfahren zur Herstellung von Wirbelschichtgranulaten aus Milchbruchmassen und ihre Verwendung in Lebensmitteln [Process for manufacture of fluidized bed granulates from milk curd, and their use in foods], German Democratic Republic Patent, DD271840, 1989.

Markvart, M., Vanecek, V. and Drbohlav, R., The drying of low melting point substances and evaporation of solutions in fluidized beds, *Br. Chem. Eng.*, **7** (1962) 503–507.

Masters, K., Recent developments in spray drying, in: Thorne, S., (ed.), Developments in food preservation, volume 2, Elsevier, London, 1983, 95–121.

Masters, K., Spray drying handbook, 5th ed., Longman, Harlow, 1991.

Mathur, K.B. and Epstein, N., Spouted beds, Academic Press, New York, 1974.

Meissner, H.P. and Mickley, H.S., Removal of mists from air in a fluidized bed, *Ind. Eng. Chem.*, **41** (1949) 1238–1242.

Metheny, D.E. and Vance, S.W., Control of particle size, *Chem. Eng. Prog.*, **58** (6) (1962) 45–48.

Nalimov, S.P., Todes, O.M. and Radin, S.I., Mechanism of thermal subdivision of granules during drying of solutions in a fluidized bed, *J. Appl. Chem. USSR*, **48** (1975) 2073–2077.

Newitt, D.M. and Conway-Jones, J.M., A contribution to the theory and practice of granulation, *Trans. Inst. Chem. Engrs.*, **36** (1958) 422–442.

Nienow, A.W. and Rowe, P.N., Fluid bed granulation, Separation Process Services, Harwell, 1975.

Nikolaev, P.I., Lyandres, S.E. and Vologodskii, L.B., Effect of gas velocity on the capacity of apparatus for drying of solutions in a fluidised bed of inert material, *Int. Chem. Eng.*, **15** (1975) 95–96.

Ormos, Z., Studies on granulation in fluidized bed, I. Methods for testing the physical properties of granulates, *Hungarian J. Ind. Chem.*, **1** (1973) 207–228.

Ormos, Z., Pataki, K. and Csukas, B., Studies on granulation in fluidized bed, II. The effect of the amount of binder on the physical properties of granules formed in a fluidized bed, *Hungarian J. Ind. Chem.*, **1** (1973a) 307–328.

Ormos, Z., Pataki, K. and Csukas, B., Studies on granulation in fluidized bed, III. Calculation of the feed rate of granulating liquid, *Hungarian J. Ind. Chem.*, **1** (1973b) 463–474.

Ormos, Z., Pataki, K. and Csukas, B., Studies on granulation in fluidized bed, IV. Effects of the characteristic of the fluidized bed, the atomization and the air distributor on the physical properties of granulates, *Hungarian J. Ind. Chem.*, **1** (1973c) 475–492.

Ormos, Z., Csukas, B. and Pataki, K., Granulation and coating in a multicell fluidised bed, in: Keairns, D.L., (ed.), Fluidization technology volume 2. Proceedings of an International Conference on Fluidization, Pacific Grove, California, 1975. Hemisphere, New York, 1976, 545–554.

Otero, A.R. and Garcia, V.G., Cake formation in a fluidised bed calciner, *Chem. Eng. Prog. Symp. Ser.*, **66** (105) (1970) 267–276.

Perez, F. and Westergaard, V., Dos Pinos extends milk powder production capacity, *Scandinavian Dairy Information*, **9** (2) (1995) 18–19.

Peterreins, F. and Kamil, G., Fluidized bed technology in brewing, German Federal Republic Patent DE20113967U1, 2002a.

Peterreins, F. and Kamil, G., Use of a drying and granulation process in brewing, German Federal Republic Patent DE10120979A1, 2002b.

Pietsch, W.B., Adhesion and agglomeration of solids during storage, flow and handling, *J. Eng. Ind. (Trans. A.S.M.E.)*, **May** (1969a) 435–449.

Pietsch, W.B., The strength of agglomerates bound by salt bridges, *Can. J. Chem. Eng.*, **47** (1969b) 403–409.

Piotrowski, A. and Kamil, G., Use of beer wort granulate, on the example of alcohol-free beer mixed drinks, *Brauwelt*, **142** (2002) 651–653.

Randolph, A.D. and Larson, M.A., Theory of particulate processes, Academic Press, London, 1971.

Rankell, A.S., Scott, M.W., Lieberman, H.A., Chow, F.S. and Battista, J.V., Continuous production of tablet granulations in a fluidized bed, II. Operation and performance of equipment, *J. Pharm. Sci.*, **53** (1964) 320–324.

Robinson, T. and Waldie, B., Dependency of growth on granule size in a spouted bed granulator, *Trans. Inst. Chem. Engrs.*, **57** (1979) 121–127.

Rowe, P.N. and Nienow, A.W., Particle mixing and segregation in gas fluidised beds: a review, *Powder Tech.*, **15** (1976) 141–147.

Rowe, P.N., MacGillivray, H.J. and Cheesman, D.J., Gas discharge from an orifice into a gas fluidised bed, *Trans. Inst. Chem. Engrs.*, **57** (1979) 194–199.

Rumpf, H. and Schubert, H., Behaviour of agglomerates under tensile stress, *J. Chem. Eng. Japan*, **7** (1974) 294–298.

Rümpler, K. and Jacob, M., Continuous coating in fluidized bed, *Food Marketing Tech.*, **12** (1998) 41–43.

Rümpler, K. and Jacob, M., Continuous agglomeration and granulation by fluidization, *Food Marketing Tech.*, **13** (1999) 31–33.

Scott, M.W., Lieberman, H.A., Rankell, A.S. and Battista, J.V., Continuous production of tablet granulations in a fluidized bed, I. Theory and design considerations, *J. Pharm. Sci.*, **53** (1964) 314–319.

Sherrington, P.J., Granulation of sand as an aid to understanding fertiliser granulation, *Chem. Engnr.*, **July/August** (1968) 201–215.

Sherrington, P.J. and Oliver, R., Granulation, Heyden, London, 1981.

Shilton, N.C. and Niranjan, K., Fluidization and its applications to food processing, *Food Structure*, **12** (1993) 199–215.

Shuhama, I.K., Aguiar, M.L., Oliveira, W.P. and Freitas, L.A.P., Experimental production of annatto powders in a spouted bed dryer, *J. Food Eng.*, **59** (2003) 93–97.

Singiser, R.E., Heiser, A.L. and Prillig, E.B., Air suspension tablet coating, *Chem. Eng. Prog.*, **62** (6) (1966) 109–115.

Smith, P.G., A study of fluidised bed granulation, PhD thesis, University of London, 1980.

Smith, P.G., Introduction to food process engineering, Kluwer, New York, 2003.

Smith, P.G. and Nienow, A.W., Growth mechanisms in fluidised bed granulation. Proceedings of an International Symposium on Particle Technology, I.Chem.E. Symp. Ser. No. 63 (1981) D2/K/1.

Smith, P.G. and Nienow, A.W., On atomising a liquid into a gas fluidised bed, *Chem. Eng. Sci.*, **37** (1982) 950–954.

Smith, P.G. and Nienow, A.W., Particle growth mechanisms in fluidised bed granulation, part I. The effect of process variables, *Chem. Eng. Sci.*, **38** (1983a) 1223–1231.

Smith, P.G. and Nienow, A.W., Particle growth mechanisms in fluidised bed granulation, part II: comparison of experimental data with growth models, *Chem. Eng. Sci.*, **38** (1983b) 1233–1240.

Sommer, K., Palzer, S., Niederreiter, G. and Wenisch, J., Process and equipment for enrichment of powders with fat components, German Federal Republic Patent, DE10055317A1, 2002.

Szentmarjay, T. and Pallai, E., Drying of suspensions in a modified spouted bed drier with an inert packing, *Drying Tech.*, **7** (1989) 523–536.

Szentmarjay, T, Pallai, E. and Regenyi, Z., Short-time drying of heat-sensitive, biologically active pulps and pastes, *Drying Tech.*, **14** (1996), 2091–2115.

Teunou, E. and Poncelet, D., Batch and continuous fluid bed coating: review and state of the art, *J. Food. Eng.*, **53** (2002) 325–340.

Tia, S., Tangsatitkulchai, C. and Dumronglaohapun, P., Continuous drying of slurry in a jet spouted bed, *Drying Tech.*, **13** (1995) 1825–1840.

Tsujimoto, H., Yokoyama, T. and Sekiguchi, I., The characterization of micro-granules produced by a tumbling fluidized bed granulator with an opposed pulsed jet assembly, *J. Soc. Powder Tech. Japan*, **35** (1998) 256–264.

Uemaki, O. and Mathur, K.B., Granulation of ammonium sulphate fertilizer in a spouted bed, *Ind. Eng. Chem. (Proc. Des. Dev.)*, **15** (1976) 504–508.

Vance, S.W. and Lang, R.W., The versatility of fluidised bed driers, *Chem. Eng. Prog.*, **66** (7) (1970) 92–93.

Walter, U. and Rümpler, K., Design and construction of a fluid bed granulator/dryer for feed additives, *Food Marketing Tech.*, **14** (2000) 48–50.

Watano, S., Yeh, N. and Miyanami, K., Drying of granules in agitation fluidized bed, *J. Chem. Eng. Japan*, **31** (1998) 908–913.

Winkelmann, M., Roebert, L., Arndt, T. *et al.*, Process for candying of coffee beans, German Federal Republic Patent, DE19902786C1, 2000.

Wurster, D.E., Air-suspension technique of coating drug particles; a preliminary report, *J. Am. Pharm. Assoc.*, **48** (1959) 451–454.

Wurster, D.E., Preparation of compressed tablet granulations by the air-suspension technique II, *J. Am. Pharm. Assoc.*, **49** (1960) 82–84.

Yu, C.Y., Xu, Y.K. and Wang, X.Z., Study of fluidized-bed spray granulation, *Drying Tech.*, **17** (1999) 1893–1904.

Zelenskaya, L.D. and Pilipenko, L.N., Fluidized-bed drying of potato puree with milk, *Izvestiya Vysshikh Uchebnykh Zavedenii, Pishchevaya Tekhnologiya*, **1** (1989) 136–138.

Chapter 6
Gas-Solid Fluidized Bed Fermentation

Principles of fluidized bed fermentation

Solid-state fermentation is an alternative approach to submerged fermentation techniques. In submerged fermentations the substrate and micro-organisms can be regarded as being homogeneously distributed in a free aqueous phase whereas in solid-state fermentations the substrate and micro-organisms are distributed pseudo-homogeneously or heterogeneously with only sufficient water available to support the metabolic activity of the fermenting micro-organisms (Bauer and Kunz, 1987). Fermenter geometries include simple trays, rotating drums, Z-blade mixers and packed beds (Ellis *et al.*, 1994). Sato *et al.* (1985) used a packed bed reactor for ethanol production from a variety of starch sources (rice, maize, sweet potato and cassava) with inert gas circulation for ethanol stripping. However, the process was slow, taking over two weeks to obtain 80% of the theoretical yield of ethanol, and the authors suggested that productivity might be enhanced by some form of solids agitation.

For ethanol production a substantial improvement over the packed bed reactor is offered by the gas-solid fluidized bed fermenter which uses a liquid instead of a solid substrate and replaces the packed bed of substrate or micro-organisms with a fluidized bed of yeast pellets (Smith *et al.*, 1997). This system can be viewed as a standard fermentation against which the performance of different bioreactors may be assessed.

All novel bioreactors attempt to increase productivity by increasing yeast cell concentration or by reducing ethanol inhibition and much work has been reported on ethanol production in fluidized beds by Bauer, Hayes, Moebus, Rottenbacher, Teuber and their co-workers; aqueous glucose solutions can be atomised within a bed of yeast particles with the latent heat for vaporisation of both ethanol and extraneous water being supplied in the inlet gas. The fluidized bed allows a uniform temperature in the fermenter and the easy addition and removal of heat, substrate, nutrients and gases, avoiding the usual high dilution of micro-organisms. Pure ethanol is condensed directly from the exhaust gases of the fluidized bed without the need for costly downstream processing whilst the removal of product in this way reduces the inhibition of the yeast by the ethanol (Smith *et al.*, 1996). High yeast cell concentrations of up to $160 \, \text{g} \, \text{l}^{-1}$ (dry weight) (Bauer, 1984) can be obtained in the reactor, comparable to the concentrations obtained in continuous stirred tank reactors (CSTR) operated with membrane recycle, i.e. $120–200 \, \text{g} \, \text{l}^{-1}$ (dry weight) (Warren *et al.*, 1994). However, gas-solid fluidized bed fermenters have two advantages over CSTRs with membrane recycle: gas stripping of ethanol from the bed, which results in substantial relief from ethanol inhibition, and the production of substrate-free ethanol which simplifies downstream processing.

Fluidization gives better heat and mass transfer than conventional solid-state methods so the temperature of fermentation is more easily controlled and more uniform (Bauer, 1984). In submerged fermentation, biological waste heat often creates a cooling problem but in the gas-solid fluidized bed fermenter this heat assists the evaporation of ethanol from the bed (Moebus *et al.*, 1981). Despite the improved mass transfer, Rottenbacher *et al.* (1985b) suggested that there are diffusional limitations within the yeast pellets. Ethanol inhibition does occur in gas-solid fluidized bed fermenters but only when inert gas is used to strip ethanol from the bed and the ethanol is only partially recovered from the off-gas (Beck and Bauer, 1989) before it is returned to the bed. One of the biggest difficulties is the initiation of fluidization of the moist yeast pellets and the maintenance of fluidization against the flow of liquid substrate (Grabowski *et al.*, 1997); flow aids such as silicic acid have been used to improve flowability (Egerer *et al.*, 1985). During operation of the bed there is a tendency for the yeast pellets to agglomerate and the bed to quench which limits the possible flow rate of substrate into the bed (Rottenbacher *et al.*, 1985b). A complicating factor is the need to maintain the moisture content of the yeast pellets above the 55% (wwb) minimum required for adequate fermentation of the substrate (Bauer, 1986).

A theoretical model (Beck and Bauer, 1989), based on ethanol inhibition alone as the limiting factor in gas-solid fluidized bed fermenters run with recirculating inert gas, suggested that the potential of this technique has not been fully explored. Hayes (1998) suggested significant improvements to the model and provided experimental confirmation of its validity.

Other than ethanol production, both Moebus and Teuber (1984) and Bauer and Kirk (1985) have investigated glutathione production from

S. cerevisiae; it is usually produced by submerged culture at low yield. Bauer and Kirk (1985) demonstrated that yield could be increased by a factor of 5–10 in a gas-solid fluidized bed. A pilot-scale air-fluidized bed has been used for the cultivation of *Aspergillus sojae* on fluidized wheat bran (Akao and Okamoto, 1983). The enzymes in *Aspergillus sojae* are used in soy sauce manufacture and gas-solid fluidized bed culture increases enzyme activity 5–15 times over traditional culture methods. Similar findings have been reported with other micro-organisms cultivated by this technique (Tanaka *et al.*, 1986). Immobilised bacteria in gas-solid fluidized beds have also been investigated for the production of ethylene oxide from ethylene (de Bont and van Ginkel, 1983) and the production of propylene oxide from propylene (Hou, 1984). In a review of solid-state fermentation in the food industry, Bauer and Kunz (1987) suggested that fluidized micro-organisms could serve as a biofilter for gases. Smith *et al.* (1997) suggested that gas-solid fluidized beds can be used to manufacture a range of volatile products, including fusel alcohols and flavour compounds, using a variety of micro-organisms.

Fermentation of glucose by *Saccharomyces cerevisiae*

Metabolism

Saccharomyces cerevisiae is able to metabolise glucose under aerobic or anaerobic conditions to provide both energy (as ATP) and biosynthetic intermediates for cell growth and maintenance (Ratledge, 1987). Under anaerobic conditions *S. cerevisiae* ferments glucose to ethanol, and under aerobic conditions both fermentation and respiration of glucose may occur. During aerobic growth respiration accounts for less than 10% of glucose consumed (Gancedo and Serrano, 1989) and the fact that *S. cerevisiae* is an efficient converter of glucose (and some other sugars) to ethanol has been exploited by the brewing, potable ethanol and fuel ethanol industries.

Since more ATP is produced by respiration of glucose than by fermentation, and since the ATP requirement for biosynthesis of cell mass is the same, it follows that to obtain the same cell yield from glucose, the yeast should consume less sugar under aerobic conditions than under anaerobic conditions, with a resultant decrease in glycolytic flux (Berry, 1982). These phenomena are referred to as the Pasteur effect. Although this effect is observed in some yeasts, in *S. cerevisiae* it is either absent (Gancedo and Serrano, 1989) or observed only under certain nutrient-limited conditions. The main reason for the absence of the Pasteur effect is that even under aerobic conditions, fermentation is still the main catabolic route for the utilisation of glucose because of the Crabtree effect (Walker, 1994). The Crabtree effect is the repression of aerobic respiration in favour of fermentation when the growth medium contains a high concentration of a readily assimilated sugar such as glucose (Berry, 1982). The threshold glucose concentration at which glucose is converted to ethanol and carbon dioxide in the presence of excess oxygen depends upon the yeast strain and environmental factors. For baker's yeast (*S. cerevisiae*) the threshold glucose concentration lies in the range 35–280 mg l^{-1} (Stear, 1990).

Production of cell mass and ethanol yield

For cell growth to occur the medium must supply carbon, nitrogen, oxygen and hydrogen for the major cell constituents and potassium, phosphorus, sulphur and magnesium for the minor cell components (Maiorella *et al.*, 1981). Vitamin requirements depend upon the yeast strain and the minor metals required for good yeast growth are calcium, zinc, iron and copper (Berry, 1982). In aerobic continuous submerged culture, a cell yield of about 0.5 g cell dry weight per g of glucose consumed is obtained at dilution rates too low to permit the operation of the Crabtree effect, thus permitting complete respiration of glucose (Fiechter *et al.*, 1987). In the industrial production of baker's yeast using molasses and ammonium salts, a batch incremental-feeding method ensures that sugar is supplied at a concentration of less than 0.1% in order to avoid the Crabtree effect. This gives a cell yield of 0.5 based on the mass of sugar consumed (Stear, 1990). Regulation of feed rate has been used when growing baker's yeast in aerobic gas-solid fluidized bed fermenters from media containing sucrose or glucose as the carbon source, since high feed rates permit operation of the Crabtree effect which leads to ethanol production under aerobic conditions. Because ethanol is stripped from the bed by the fluidizing gas, cell yields may be reduced (Moebus *et al.*, 1981; Mishra *et al.*, 1982).

The manufacture of potable and industrial ethanol requires that as little substrate as possible is diverted into production of new cell mass. In theory, 1 g of glucose should yield 0.511 g of ethanol and 0.489 g of carbon dioxide but in practice, in batch aerobic fermentations, it is usually found that about 0.1 g of the glucose is diverted into new cell mass such that (Maiorella *et al.*, 1981)

$$\text{Glucose (1g)} \rightarrow \text{Ethanol (0.46g)} + \text{Carbon dioxide (0.44g)}$$
$$+ \text{New cells (0.1g)}$$

More glucose can be converted to ethanol if cell recycle is used (Jackman, 1987) or by excluding a nitrogen source from the fermentation medium and using anaerobic conditions so that produced ethanol cannot be reconsumed. This principle was applied by Moebus and

Table 6.1   Expected yields from fermentation of glucose.

Product	Yield (w/w)
Ethanol	0.45–0.49
Carbon dioxide	0.43–0.47
Glycerol	0.02–0.05
Succinate	0.005–0.015
Fusel oil	0.002–0.006
Acetate	0–0.014
Butylene glycol	0.002–0.006
Cell material	0.007–0.017

Teuber (1982a) and Rottenbacher (1985) to gas-solid fluidized bed fermentation of glucose to ethanol. Whilst the feed solutions used by Moebus and Teuber (1982a) contained only glucose, Rottenbacher (1985) included potassium dihydrogen phosphate and magnesium sulphate as these salts are beneficial to fermentation (Jones and Greenfield, 1984) and obtained higher yields of ethanol (0.475–0.511) than Moebus and Teuber (1982a) (0.43–0.47). These yields are similar to those obtained by submerged fermentation as shown in Table 6.1. The table also lists the main by-products of fermentation; examples of fusel oils are n-propanol, isobutanol, amyl alcohol and isoamyl alcohol (Ingledew, 1993).

Factors affecting ethanol production

A summary of the factors which are known to influence ethanol production from glucose in a gas-solid fluidized bed fermenter, or which may have an influence based on observations with submerged fermentations, is shown in Figure 6.1. In anaerobic beds, the key factors are the fermentation temperature and ethanol inhibition, both of which have a dramatic effect on the specific rate of ethanol production. Bed dehydration and its influence on yeast pellet moisture content is also important, since a failure of fermentation may occur if the pellets become too dry (Bauer, 1986).

Glucose concentration

At glucose concentrations between 3.5 and $150\,g\,l^{-1}$ the specific rate of ethanol production is essentially at a maximum value and beyond this range substrate inhibition of growth and fermentation can occur (Maiorella *et al.*, 1981). In contrast to submerged fermentations, Moebus and Teuber (1982a) supplied a glucose solution at $469\,g\,l^{-1}$ to a fluidized

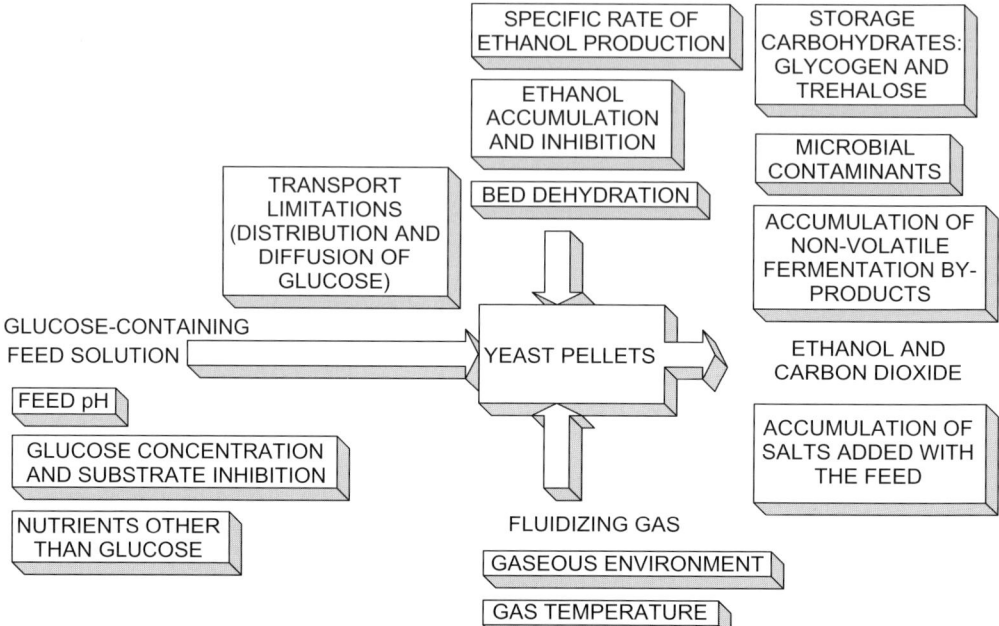

Figure 6.1   Factors affecting ethanol production from glucose using baker's yeast (*S. cerevisiae*) in a gas-solid fluidized bed fermenter. Reproduced from Hayes (1998) with permission.

bed of baker's yeast pellets with no apparent decrease in yeast cell viability over a 7.7 hour fermentation period. Rottenbacher (1985) assumed that the residual glucose was in solution and located only in the intraparticle pores of the yeast pellets and the mean glucose concentration in this bed was only about $30\,\mathrm{g\,l^{-1}}$. This estimated value is considerably less than both the feed concentration and that suggested by Maiorella *et al.* (1981) for substrate inhibition. This suggests that glucose inhibition of fermentation in gas-solid fluidized beds is unlikely to be important unless some other factor intervenes to cause glucose to accumulate within the pellets or on their surface.

Gaseous environment

The two inert gases used in anaerobic gas-solid fluidized bed fermenters are carbon dioxide (Moebus and Teuber, 1982a) and nitrogen (Rottenbacher, 1985). The former found that the addition of air at the beginning of the run substantially increased carbon dioxide production compared to a bed run under pure carbon dioxide. An identical bed operated under strict anaerobiosis appeared to ferment glucose slowly at the beginning of the run, with a gradual rise in carbon dioxide production as the run progressed. Moebus *et al.* (1984) explained that under strictly anaerobic conditions early in the run insufficient NADH was being re-oxidised to NAD$^+$ by fermentation and that this problem was overcome by adding air (oxygen) to enable NAD$^+$ to be regenerated oxidatively and speed up the rate of fermentation at the expense of some glucose. This explanation assumes that the Crabtree effect can be lifted in order for the oxidation of glucose to proceed. Using nitrogen as the fluidizing gas, Rottenbacher (1985) did not observe the same slow fermentation and it was not necessary to add air to speed up the rate of ethanol production. This suggests that it may be advantageous to operate anaerobic beds with nitrogen, rather than carbon dioxide. The fact that the choice of inert gas had no apparent effect on the rate of ethanol production in submerged culture (Thibault *et al.*, 1987) also suggests that there may be an important difference between gas-solid fluidized bed fermentations and submerged culture.

Ethanol inhibition

Ethanol production generally ceases when the ethanol concentration in the medium reaches $110\,\mathrm{g\,l^{-1}}$ because ethanol itself has a toxic effect on yeast metabolism (Maiorella *et al.*, 1981) and the main site of toxicity is the cellular membrane and membrane transport processes (Pamment *et al.*, 1990). Cell growth is more easily inhibited by ethanol than is the specific rate of ethanol production (Pamment, 1989). There is a large difference in toxicity between ethanol which has been produced by the fermenting yeast *in situ* by fermentation, and ethanol from an external source that may be added to the fermenting medium (Pamment, 1989).

Temperature

Baker's yeast has an optimum growth temperature of about 35°C and an optimum fermentation temperature of 40°C (Burrows, 1970). The temperature range for growth of *S. cerevisiae* is about 3–45°C and at about 50°C or greater there is a rapid decline in cell population (van Uden, 1989). The toxic effects of ethanol are enhanced at high fermentation temperatures, resulting in a reduction in the maximum temperature for cell growth and an enhancement of thermal death (van Uden, 1989). Ethanol yield tends to be higher at low temperature in batch fermentations (Casey and Ingledew, 1986).

pH

The absolute pH range for the growth of most yeasts is 2.4–8.6 with the optimum being about 4.5 (Atkinson and Mavituna, 1983). The specific rate of ethanol production is very sensitive to pH values above 5 and at pH 6 it is only 50% of the maximum value (Jones and Greenfield, 1984). With increasing pH there is a reduction in the yield of ethanol from glucose and a rise in the yields of both glycerol and acetic acid.

Moisture content

Bauer (1986) recommended that the moisture content of pressed baker's yeast pellets should be maintained between 55% and 65% (wwb) for satisfactory fermentation of the feed glucose. The original moisture content of the yeast is about 70% (wwb) and any decline in moisture content in the bed is due to a mismatch between the drying rate in the bed and the rate at which water is added with the feed solution. Such a mismatch occurred in work reported by Moebus and Teuber (1984) who obtained ethanol yields of 0.36 and 0.21 at final yeast moisture contents of 55.9% and 32.6% respectively.

Mass transport limitations

The specific rate of ethanol production from glucose is usually expressed as g of ethanol produced per hour per g of yeast cell dry weight ($g\,gDW^{-1}h^{-1}$). For the baker's yeast used in work by Hayes and co-workers (Hayes, 1998; Smith *et al.*, 1996, 1997) and reported here, Maynard (1993) obtained a specific rate of ethanol production in submerged culture of just over $0.3\,g\,gDW^{-1}h^{-1}$ in a batch fermentation at 30°C. However, baker's yeast strains vary in their capacity to produce ethanol from glucose. Rottenbacher *et al.* (1985a) obtained $0.75\,g\,gDW^{-1}h^{-1}$ with another commercial baker's yeast (Deutsche Hefewerk, strain DHW DZ) in submerged culture in a batch fermenter at 30°C. Specific rates of ethanol production by baker's yeast pellets in gas-solid fluidized bed fermenters are much lower (50% or more) than those obtained in submerged culture because of diffusional limitations in the pellet and difficulties in adequately distributing the glucose feed solution to the whole of the fluidized bed (Bauer, 1986). Glucose applied to the surface of the yeast pellets diffuses into the pellet through the intraparticle pores not occupied by yeast cells (Rottenbacher, 1985) because glucose does not permeate freely through the lipid membranes of the yeast cells. The simultaneous diffusion and consumption of glucose gives rise to a glucose concentration profile along the pellet radius which may, depending upon the operating conditions of the fluidized bed, cover only a fraction of the whole pellet volume (Rottenbacher, 1985). Increasing solids mixing in aerobic gas-solid fluidized bed fermenters has been demonstrated as a method of better distributing the glucose feed solution so that the risk of agglomeration is reduced. Better solids mixing also makes it possible to increase the mass flow rate of glucose into the bed which raises the specific rate of ethanol production (Rottenbacher *et al.*, 1985b, 1987).

Fluidized bed fermentation systems

Basic fluidization considerations

Gas-solid fermenters are operated in the bubbling bed regime (Rottenbacher, 1985) and the yeast pellets fall into Geldart groups B and D. If the bed contains fines, they will be elutriated from the bed if the superficial gas velocity equals or exceeds their terminal falling velocity. For fluidized beds of yeast pellets, it has been found necessary either to install a filter above the freeboard of the bed (Rottenbacher, 1985) or to pass the off-gas through a cyclone (Moebus *et al.*, 1981; Mishra *et al.*, 1982) to maintain an inventory of the yeast mass. Superficial gas velocities greater than $0.45\,\mathrm{m\,s^{-1}}$ (Mishra *et al.*, 1982) but below $1.3\,\mathrm{m\,s^{-1}}$ (Bauer, 1985) have been employed. Different workers have used different pellet sizes, with Bauer and co-workers preferring to use $800\,\mu\mathrm{m}$ diameter extruded pellets in small diameter (0.2 m) beds whilst others have used grated pellets. Mishra *et al.* (1982) used grated pellets of a wide size distribution and mean size $943\,\mu\mathrm{m}$ in a 0.192 m diameter bed whilst Moebus and co-workers have generally used grated pellets in the size range 1–2 mm in a 0.19 m diameter bed (Moebus and Teuber, 1982a, b). Hayes and co-workers (Smith *et al.*, 1996, 1997; Hayes, 1998) used both 1 mm grated and 1 mm and 2 mm extruded yeast pellets.

Whether baker's yeast pellets are simply to be dried for active dried yeast (ADY) manufacture or used for ethanol production, there is some initial difficulty in getting the pellets to fluidize. This is because the outer surface of the pellet is covered with a liquid layer which creates liquid bonds between adjacent yeast pellets. This reduces the viscous drag force exerted by the fluidizing gas on the bed and channelling of the gas occurs (Egerer *et al.*, 1985). It becomes less difficult to fluidize the yeast pellets as the initial moisture content of about 70% (wwb) is reduced by drying to 35% (wwb) (Grabowski *et al.*, 1997). Egerer *et al.*

(1985) investigated the effect on minimum fluidizing velocity of adding a hydrophobic flow aid (silicic acid) at about 1% by mass of yeast to the surface of extruded, cylindrical yeast pellets. They found that flow aid greatly improved the fluidization characteristics of the yeast pellets up to a moisture content of 60% (wwb). At higher moisture contents there was a significant rise in minimum fluidizing velocity but this value was still lower than that for untreated pellets.

Anaerobic ethanol production

Since anaerobic conditions give fermentation, rather than respiration of glucose, Moebus and Teuber (1982a) examined ethanol production from simple glucose solutions using grated baker's yeast pellets, 1–2 mm in diameter, fluidized by carbon dioxide; the gas was recirculated inside a gas-tight closed circuit and the excess carbon dioxide produced by fermentation was vented and a water-cooled partial condenser was used to recover ethanol and water vapour from the system. They used a bed of diameter 0.19 m fitted with a rotating steel cone agitator at the base of the bed, apex uppermost, and an annular gap of 1.5–2 mm at the bed wall to create a gas jet in order to control agglomeration. The superficial gas velocity was approximately $0.7\,m\,s^{-1}$. Heat generated by the blower and biological heat of fermentation appear to have been sufficient to sustain the fermentation temperature at 17.5–19°C which gave a maximum rate of ethanol production of $49.7\,g\,h^{-1}$ from a $469\,g\,l^{-1}$ glucose solution using a 3 kg bed of yeast. This value was raised to $86.4\,g\,h^{-1}$ with the aid of a wet gas scrubber to assist the water-cooled partial condenser in recovering ethanol and therefore reducing ethanol inhibition in the bed. The scrubber was responsible for a 3–5 fold reduction in the ethanol accumulated in the water phase of the yeast cells in the bed.

The production of ethanol from cooked rice starch (Moebus and Teuber, 1985) differs from the normal process of spraying the carbon source into the bed since all of the carbon source is made available at the start of the run, subject only to the breakdown of starch to glucose (and maltose) by amylases. The starch (0.3 mm particles), amylases and yeast pellets were mixed in the bed and water sprayed in to maintain the fermentative activity of the yeast. The fermentation was carried out at 31.5°C.

Bauer and co-workers (Rottenbacher, 1985; Rottenbacher *et al.*, 1987) carried out anaerobic fermentation under nitrogen gas at 18–20°C in a gas-tight closed circuit with a partial condenser followed by a gas adsorber for ethanol and water vapour recovery. The highest rate of ethanol production was about $50\,g\,h^{-1}$ with a bed of 2.4 kg of extruded cylindrical baker's yeast pellets of diameter $800\,\mu m$. This rate is less than that achieved by Moebus and co-workers with a partial condenser followed by a wet gas scrubber but similar to that with a partial condenser alone. The wet gas scrubber, activated carbon filter and adsorber are all aids to reducing ethanol inhibition in an anaerobic system where a partial condenser is used for ethanol recovery. Unfortunately, they create additional problems of ethanol recovery which probably outweigh any economic benefit arising from the reduction in ethanol inhibition. According to Moebus and Teuber (1985), use of a wet gas scrubber reduces the recovered ethanol concentration from 9.1% (v/v) to only 0.3% (v/v), and recovery of ethanol from an activated carbon filter is a time-consuming process.

Aerobic ethanol production

Moebus *et al.* (1981) and Mishra *et al.* (1982) showed that adding excess glucose to a bed of yeast pellets fluidized with air gave rise to aerobic ethanol production. Although the ethanol yield in the aerobic system was lower (at about 0.1) than in an anaerobic system (about 0.43–0.47), Bauer and co-workers considered that ethanol inhibition would be negligible since fresh air rather than recirculated inert gas laden with ethanol vapour could be used to strip ethanol from the bed. Furthermore, there was the possibility of approaching the anaerobic yield of ethanol if sufficient glucose could be added to the bed without quenching. Consequently, Bauer and Rottenbacher (1984) used a fluidized bed without an agitator, albeit with an optoelectronic monitoring system to control agglomeration in the bed which periodically shut off the feed, allowing the bed to recover. Under aerobic conditions at 22°C Bauer (1985) was able to obtain an ethanol production rate of $65\,g\,h^{-1}$ with a corresponding ethanol yield of 0.37, about 70% of the theoretical maximum.

Bauer (1985) reported the scale-up of a gas-solid fluidized bed fermenter from approximately 0.2m in diameter with 3kg of yeast to a 0.55m diameter bed and 20kg with a simultaneous reduction in particle size from $800\,\mu m$ to $500\,\mu m$ in an attempt to improve the vertical mixing in the bed. Increasing the bed temperature to 34°C in the 0.55m bed reduced the ethanol yield from glucose to 0.33 but this was more than offset by the specific rate of ethanol production which reached a maximum of $0.257\,g\,gDW^{-1}h^{-1}$. This corresponded to an ethanol productivity of about $40\,g\,l^{-1}h^{-1}$ where the reaction volume is defined as the volume occupied at minimum fluidizing conditions by the yeast pellets plus the bed voidage. The effect of increasing the bed diameter, reducing the pellet diameter and increasing the bed temperature was to approximately treble the rate at which glucose could be supplied per unit mass of yeast in the bed from 78 to $230–240\,g\,h^{-1}$ per kg of 'wet'

yeast. Superficial gas velocities were the same in each case at
1.2–1.3 m s^{-1}.

Agglomeration, quenching and the glucose sink

The physics of gas-solid fluidized bed fermentation has a great deal in
common with fluidized bed granulation (see Chapter 5). Of course, in
gas-solid fluidized bed fermentation it is desirable that no particle
growth occurs and that glucose, a potential binder in granulation
terms, is fermented instead to ethanol. Wet quenching results if the
overall material and energy balance in the bed is not met and excess
liquid remains on the surface of the pellets. As in granulation, even if
the overall material and energy balances are satisfied, localised wet
quenching can arise if the feed solution is poorly distributed because
of either poor feed atomisation or inadequate solids circulation in the
bed. Thus atomisation with a twin-fluid nozzle is preferred (Moebus
et al., 1981) either above the surface of the bubbling bed (Rottenbacher,
1985) or into the bed (Mishra *et al.*, 1982). Dry quenching is likely to
occur if glucose is supplied to the bed at a rate in excess of the capacity
of the yeast pellets to ferment it and excess glucose accumulates on the
surface of the pellets. Thus the concept of a dynamic equilibrium
between binding forces which cause agglomerates to form and disruptive forces which cause them to break up, and which is used to explain
particle growth in fluidized bed granulation, applies equally to gas-
solid beds used for fermentation reactions and increasing excess gas
velocity or initial pellet size should therefore reduce the possibility of
agglomeration.

Smith and Nienow (1983) showed that in a batch fluidized bed granulator there was a considerable 'no growth' period when a binder
solution was sprayed into a bubbling bed of porous particles followed
by a sharp rate of increase in particle size due to agglomeration; intraparticle porosities between 0.4 and 0.45 have been reported for extruded
yeast pellets (Rottenbacher, 1985; Strumillo *et al.*, 1995). In order for
glucose to diffuse into the pellet it is necessary for the pores to be filled
with liquid (Rottenbacher, 1985). In the feed zone the pellet has been
modelled as receiving a finite volume of feed solution which wets the
entire pellet surface (Rottenbacher *et al.*, 1985b). Glucose therefore
enters the pores of the yeast pellet by diffusion, driven by a concentration difference across the pellet radius, and the accumulation of glucose
within the yeast pellet or on its surface is limited by fermentation to
ethanol and carbon dioxide.

Hayes (1998) proposed that the disposal of glucose by the twin processes of diffusion and fermentation can be viewed as a 'glucose sink'
and that whilst the sink is operating there is little glucose available on

the surface of the yeast pellets for the formation of solid bridges. The glucose sink therefore acts to reduce the binding force which depends upon the size of the solid bridges. Thus, a failure of fermentation or diffusion, an oversupply of glucose or a maldistribution of glucose can all lead to uncontrolled pellet growth by agglomeration and dry quenching of the bed.

The solubility of carbon dioxide in water is very low and it can be assumed that during fermentation very little carbon dioxide accumulates in the pellet and that most is transferred to the fluidizing gas (Rottenbacher *et al.*, 1985b). Essentially, free glucose is confined to the pellet pores because it cannot diffuse freely through the lipid membranes of the yeast cells (Gancedo and Serrano, 1989). However, ethanol can freely diffuse out of the yeast cells and into the pellet pores to give a uniform concentration in the water phase of the pellet (Rottenbacher, 1985). In gas-solid fluidized bed fermenters, ethanol is stripped from the bed by the fluidizing gas, and in anaerobic systems, the ethanol concentration in the pellet rises to an equilibrium value dependent upon the prevailing ethanol-water vapour-liquid equilibrium between the pellet and the fluidizing gas (Beck and Bauer, 1989).

In a model for aerobic ethanol production, Rottenbacher (1985) proposed that a glucose concentration profile develops along the pellet radius. The surface glucose concentration oscillates at a frequency set by the mean pellet circulation time which for small diameter beds ($0.2\,m$) is in the range of 5–10 seconds. The decline in glucose concentration at the pellet surface is rapid and accomplished largely in the first second after the pellet is contacted with glucose solution. Thus for a population of pellets there will be a mean aqueous glucose concentration in the bed which increases with increased glucose feed rate or shorter circulation time. Four categories of glucose can be distinguished:

(1) solid glucose on the surface of the yeast pellet similar to that found with layered growth and arising from broken embryonic agglomerates
(2) solid glucose existing as solid bridges within agglomerates
(3) aqueous glucose on the surface of the yeast pellets; and
(4) aqueous glucose in the pellet void volume.

Moebus and Teuber (1985) suggested that solid glucose bridges, broken or intact, are themselves fermentable given sufficient available moisture and sufficient time.

At the start-up of aerobic fermentation, ethanol vapour and carbon dioxide production begins almost immediately glucose is applied to the bed (Rottenbacher *et al.*, 1987), climbing to peak volumetric levels in the off-gas within about 15 minutes of applying the feed solution. The levels then decline slightly and stabilise at about 30 minutes run time. A similar pattern of ethanol vapour production is observed in anaerobic fermentation (Bauer and Rottenbacher, 1985) but the rise in the partial vapour pressure of ethanol vapour in the off-gas occurs more slowly due to ethanol accumulation in the bed. Operating within the capacity of the glucose sink in an aerobic fermenter, Bauer (1984) observed that it took about 15 minutes for the rate of ethanol production to climb to its new value following a step increase in the glucose feed rate. This indicates that the glucose sink has 'inertia'.

Moebus and Teuber (1982b) demonstrated in an anaerobic fluidized bed fermenter that exceeding the maximum capacity of the glucose sink greatly impairs the fluidization of the bed. Having established that the maximum rate of glucose addition was $212.7\,g\,h^{-1}$, the experiment was repeated at a feed rate of $150\,g\,h^{-1}$ for two hours followed by a step increase to $235\,g\,h^{-1}$ which was sustained for only 10 minutes. At this point the bed dry quenched which was attributed to an accumulation of glucose in the bed. The rate of ethanol production peaked a few minutes after the supply of glucose had been shut off, indicating that accumulated glucose was being fermented and that some of the inertia in the glucose sink at $150\,g\,h^{-1}$ had been overcome. The bed began to dry quench because the maximum capacity of the glucose sink had been exceeded but the rapidity of quenching was due in part to the slow response of the system to a step increase in the glucose feed rate (i.e. inertia in the glucose sink).

The factors affecting the capacity of the glucose sink can be divided into three groups. First, those affecting the distribution of glucose, i.e. feed atomisation and solids mixing. For a given bed diameter and yeast mass, the glucose sink can be increased by using more than one nozzle and reducing the mean pellet circulation time by increasing the excess gas velocity. Second, those factors affecting the fermentation of glucose (principally temperature, ethanol inhibition of fermentation and pellet moisture content) which were discussed on page 191. Finally, the diffusive flux of glucose into the yeast pellet depends upon the pellet porosity and size and the feed glucose concentration. Smaller pellets allow more of the yeast pellet volume to be recruited for fermentation but increase the risk of agglomeration. Increasing the feed glucose concentration also increases the binding force at the same feed rate due to greater mass deposition of glucose on the pellet surface.

The work of Hayes (1998)

A description of the experimental system

Figure 6.2 is a schematic diagram of the anaerobic gas-solid fluidized bed fermenter used by Hayes and co-workers (Hayes, 1998; Smith

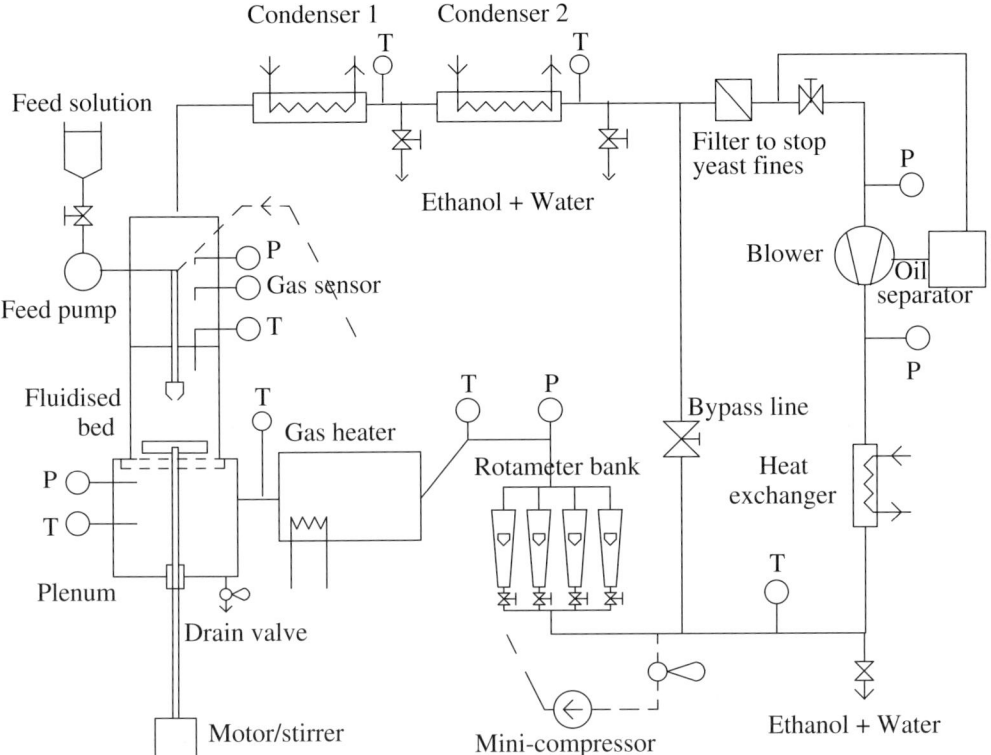

T = temperature
P = pressure

Figure 6.2   Experimental gas-solid fluidized bed fermenter rig. Reproduced from Hayes (1998) with permission.

et al., 1996, 1997). The fluidizing gas was discharged from the compressor, via the heat exchanger, rotameters, heater vessel, fluidized bed, condensers and the gas filter and returned to the blower inlet. During a fermentation run the compressor inlet pressure was usually slightly below atmospheric whilst the discharge pressure was usually about 114 kPa. Air leaving the compressor reached a temperature of about 85°C and was reduced to about 10°C in a water-cooled shell-and-tube heat exchanger (surface area approximately 4.7 m²) before passing to the rotameter bank. An atomised glucose feed solution was sprayed by the twin-fluid nozzle into the bubbling bed of yeast pellets which was fluidized with heated inert gas. The resulting fermentation generated ethanol and carbon dioxide. Evaporated ethanol and water vapour, together with gaseous carbon dioxide, were stripped from the bed and carried out with the off-gas.

Condensation of ethanol and water vapour occurred in both the two condensers and the heat exchanger. The condensed ethanol-water mixture was drawn off at 10-minute intervals during an experimental run with about 60% of the total coming from condenser 1, 30% from condenser 2 and 10% coming from the heat exchanger. Each condenser was manufactured from 1/4 inch OD copper tube with a condensing surface of approximately $0.5\,m^2$ mounted horizontally within a glass cylinder of length 0.5 m and inner diameter 0.155 m. The condensers were inclined slightly to facilitate condensate run-off. Well downstream of the second condenser, the off-gas which had passed the condensers was mixed with gas from the bypass line and returned to the compressor inlet via a filter designed to capture yeast fines. Excess gas due to fermentation was allowed to leak from the system so that the blower discharge pressure reached a steady value, and to ensure that the system did not create a safety hazard by becoming over-pressurised. The heater vessel was fitted with five 1 kW elements. The fluidizing gas temperature was measured at the exit to the heater vessel and a PID controller enabled this to be set from room temperature up to in excess of 75°C. The set temperature was usually 35°C ± 2°C using 2 kW of heating. Oxygen, carbon dioxide and ethanol vapour concentrations in the fluidizing gas were measured by on-line gas sensors at two points on the rig: in the freeboard above the bed surface and downstream of the two condensers.

The fluidized bed consisted of four main parts: a stainless steel plenum chamber, a distributor plate, a 0.155 m diameter by 1.6 m high upright glass cylinder and a motor/stirrer assembly. Early experiments carried out with a 2 mm thick sintered bronze porous distributor plate showed that the high pressure drop (about 12–17 kPa at a gas flow rate of $79\,m^3\,h^{-1}$) developed across the plate induced 'surging' by the blower. For subsequent experiments this was replaced by a plate consisting of four 0.5 mm thick aluminium sheets with 80 diamond-shaped perforations per cm^2 giving a much lower pressure drop of about 800 Pa at the same gas flow rate. The usual mass of yeast pellets was 3 kg which gave a loose-packed bed height of about 0.27 m. Experiments were carried out with a freely fluidized bed and with a gate impeller, 0.15 m in diameter and rotating at 51 rpm. The gate had four blades to each of the lower and upper parts, constructed from 3 mm thick brass plate inclined at 45° and with the leading edge of the lower closest to the distributor plate, joined by two vertical blades 325 mm in length. The drive shaft was attached to the lower part of the gate and passed through the distributor plate and the base of the plenum chamber to be driven by a motor mounted beneath the plenum chamber.

The feed system consisted of a reservoir, a plunger-type feed pump, a twin-fluid atomising nozzle (giving a round spray of angle 13° in free air) and a mini-compressor. Inert gas for the twin-fluid nozzle was drawn between the outlet of the heat exchanger and the rotameter bank by a small compressor and delivered at 60 kPa. The gas flow rate from the nozzle was about $0.8 \, m^3 h^{-1}$ and the feed rate was typically in the range 500–1400 ml h^{-1} giving a volumetric ratio (NAR) of about 570–1600. At a typical fluidizing gas flow the atomising gas added about 1% volume to the total gas flow out of the bed. The nozzle was located 0.23 m above the distributor plate so that the nozzle penetrated the bed to a depth of about 0.04 m before fluidizing gas was applied and ensured that the nozzle was always submerged when the bed was fluidized.

Hayes and co-workers used three types of yeast pellet:

(1) spherical brewer's active dried yeast (ADY) pellets
(2) 1 mm diameter grated pellets, formed in a food processor from pressed baker's yeast blocks; and
(3) extruded cylindrical pellets, with an approximate equivalent diameter of either 1 or 2 mm, formed from pressed baker's yeast blocks.

The properties are summarised in Table 6.2. All pellets made from pressed baker's yeast were found to be very cohesive. Coating the yeast pellets with a flow aid was only partially successful in reducing the problem. By progressively drying the pellets in the bed, the fluidization behaviour improved to the point where there were no stable channels and no agglomerates held together with liquid bridges.

Dry quenching experiments

In an attempt to separate the fermentation of glucose from its propensity to cause agglomeration and quenching, Hayes (1998) used ADY

Table 6.2 Properties of yeast pellets used by Hayes and co-workers (Hayes, 1998).

	1 mm extruded	2 mm extruded	Grated	Brewer's ADY
Moisture content % (wwb)	68.67	68.80	68.20	8.48
Particle density (kg m^{-3})	1100	1100	1100	1300
Surface-volume mean diameter (μm)	1055	2005	1078	1131
Measured u_{mf} (m s^{-1})	0.52	0.9–1.1	0.47	0.38
Measured bed voidage at u_{mf}	0.46	0.46	0.48	0.42

pellets because their initial moisture content was below the 20% (wwb) found by Koga *et al.* (1966) to be necessary for the metabolism of glucose. The bed temperature was maintained at about 35°C with the heat supplied to the fluidizing gas greatly in excess of that required to evaporate water at the feed rates used. When water was sprayed into the bed at 350 ml h^{-1}, large agglomerates up to 4 cm diameter were formed and the bed quenched. The smallest agglomerates (consisting of 3–4 pellets) were held together with solid bonds of dried yeast cells with individual yeast pellets being clearly distinguishable. In the large agglomerates, individual yeast pellets were distinguishable only in the outer shell whilst the pellets at the interior were fused together. Thus the accumulation of unfermented glucose is not the only mechanism that can bring about particle growth since the inherent solubility of ADY pellets in water appears to permit the formation of solid bonds from the yeast itself when the pellets are wetted in the feed zone and redried in the bulk of the bed.

Using an impeller in the bed increased the tolerance of ADY pellets to dry quenching and allowed a water feed rate of 210 ml h^{-1} to be sustained without quenching. More significantly, increasing the excess gas velocity from 0.77 m s^{-1} to 2.76 m s^{-1} also increased the tolerance of ADY pellets to quenching. Using 250 g l^{-1} aqueous glucose as the feed instead of water caused the ADY pellets to quench quickly in a bed without an impeller. However, with an impeller the bed was able to sustain an intermediate glucose concentration of 100 g l^{-1} at a low feed rate of 210 ml h^{-1} without quenching. Increasing the feed rate to 280 ml h^{-1} at the same concentration, or increasing the concentration to 250 g l^{-1} at the same feed rate, resulted in bed quenching.

Ethanol production with grated pellets

Ethanol was produced successfully using a 3 kg bed of grated yeast pellets, glucose solutions with a concentration of 103 g l^{-1} and bed temperatures of about 20°C. More ethanol was produced under anaerobic conditions, using either carbon dioxide or nitrogen as the fluidizing gas, than aerobically with air, because anaerobic conditions prevent the reconsumption of ethanol by the yeast. However, even with air as the fluidizing gas, Hayes (1998) obtained a higher ethanol yield (0.09) than Mishra *et al.* (1982) (about 0.01). The anaerobic fermentations were carried out in three distinct phases: the addition of 103 g of glucose to the bed, followed by 500 g of water to permit the fermentation of residual glucose, and finally drying with a high fluidizing gas temperature (65°C) in order to drive accumulated ethanol from the bed and into the condensate. The most successful operation was under nitrogen where the highest glucose feed rate of 45.2 g h^{-1} was sustained and the highest ethanol productivity ($15.3\,g\,h^{-1}$) was obtained; this is higher than the $14.6\,g\,h^{-1}$ obtained by Moebus and Teuber (1982a). However, the yield of ethanol was considerably lower (0.34 compared to 0.47).

Despite the use of an impeller the bed defluidized in each of these experiments. The overall water balance was satisfied at the maximum feed rate and Hayes (1998) concluded that localised wet quenching occurred because glucose solution was trapped within the interstices of large agglomerates. This glucose then fermented under conditions approximating to those found in submerged culture at concentrations sufficiently high for the operation of the Crabtree effect (i.e. aerobic ethanol production).

Ethanol production with extruded pellets

Using extruded yeast pellets rather than grated pellets proved to be far more successful and demonstrated the viability of the gas fluidized bed fermentation technique. The bed was pre-dried for 10–30 minutes to ensure good fluidization prior to the application of the feed solution. After first optimising the feed composition, the effects of both feed concentration and fluidizing gas were investigated. These variables were examined at fermentation temperatures (18.1–20.5°C) and volumetric feed rates (412–$477\,l\,h^{-1}$) similar to those used for grated pellets although important differences were that the feed solutions were applied continuously, rather than intermittently, and lower superficial gas velocities (0.98–$1.24\,m\,s^{-1}$) were employed such that elutriation was observed to be small. Conditions were established that allowed the fermenter to be operated without an impeller, and without quenching, for up to seven hours. Since the overall water balance was always satisfied, the causes of eventual quenching were either localised wet quenching or exceeding the capacity of the glucose sink. The rates of ethanol production achieved with 3 kg of yeast pellets in this 0.155 m diameter bed of about $100\,g\,h^{-1}$ were twice that of the previously reported maximum value of $49.7\,g\,h^{-1}$ achieved by Moebus and Teuber (1982a) with the same mass of yeast in a similar-sized bed (0.19 m diameter).

Figure 6.3 shows the time course of ethanol production as a function of feed composition (Table 6.3). The best feed solution was feed D which permitted the bed to run for eight hours without quenching and resulted in a higher ethanol yield (0.46) than feed A (0.35) which contained only glucose. Potassium and magnesium ions are known to stimulate fermentation (Jones and Greenfield, 1984; Maynard, 1993). Feed C caused the bed to quench after only two hours because, under anaerobic conditions, the yeast extract supplied was unable to be assimilated sufficiently quickly and the remainder was therefore

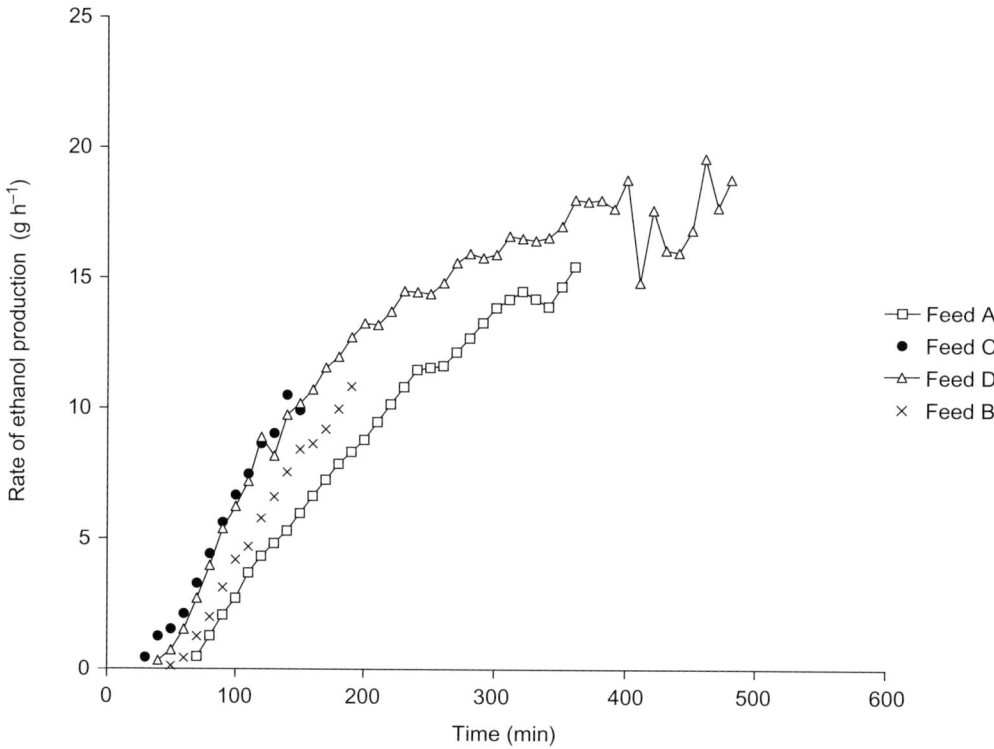

Figure 6.3 Effect of fermentation medium composition on the rate of ethanol production from glucose using extruded yeast pellets. Reproduced from Hayes (1998) with permission.

Table 6.3 The composition of glucose-based feed solutions used for anaerobic ethanol production (Hayes, 1998).

	Concentration (g l^{-1})			
	Feed A	Feed B	Feed C	Feed D
Glucose	100	100	100	100
KH$_2$PO$_4$	–	–	5	5
MgSO$_4$·7H$_2$O	–	–	0.4	0.4
(NH$_4$)$_2$·SO$_4$	–	–	3	–
Yeast extract	–	–	10	–
Citric acid monohydrate	–	8.42	–	–
Trisodium citrate dihydrate	–	14.67	–	–
pH	4.8	4.6	5.6	4.3

available to act as a particle binder. The 0.2 M citrate buffer in feed B was formulated to give the optimum pH for the fermentation of glucose and, indeed, gave a higher ethanol yield (0.40) than glucose alone. However, early quenching of the bed was observed and it is likely that

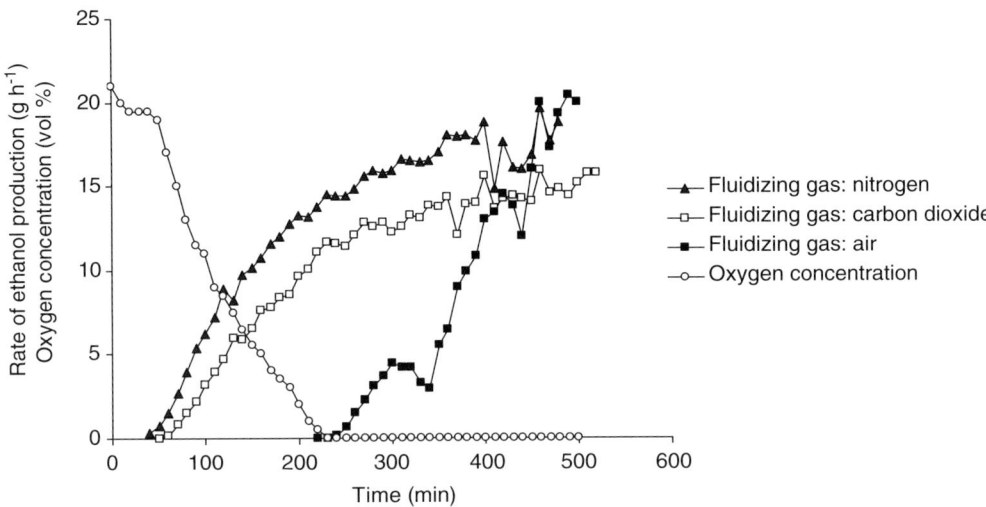

Figure 6.4   Effect of fluidizing gas composition on the rate of ethanol production from glucose using extruded yeast pellets. Reproduced from Hayes (1998) with permission.

the high concentrations of non-glucose components (citric acid monohydrate and trisodium citrate dihydrate) were responsible.

The use of nitrogen, carbon dioxide and air as fluidizing gases was compared to determine the effect on the time course of ethanol production (Figure 6.4) and on ethanol yield. Feed D was the chosen medium in each case and an impeller was used in each experiment. The fermentation under air revealed two distinct phases. Initially aerobic conditions obtained followed by anaerobic fermentation; ethanol was found in the condensate only when the oxygen initially present in the rig had been consumed. However, the use of air appeared to be responsible for severe elutriation. Hayes (1998) suggested that because more carbon dioxide is produced under aerobic conditions, and because this is generated within the pellet, the yeast pellets were more prone to disintegration from internal gas pressure with a resultant generation of fines. A higher ethanol yield (0.46) was obtained when nitrogen was the fluidizing gas compared with carbon dioxide (0.36) which was reflected in higher rates of ethanol production at the condenser. Thibault *et al.* (1987) suggested that elevated concentrations of dissolved carbon dioxide in the medium or in the yeast cells may inhibit fermentation as a result of a decrease in medium pH, or by inhibition of the enzyme system responsible for ethanol synthesis. However, in submerged culture, carbon dioxide has been observed to inhibit the fermentation of glucose by baker's yeast to a greater degree than nitrogen only at pressures greater than about 2 MPa. In other work on anaerobic

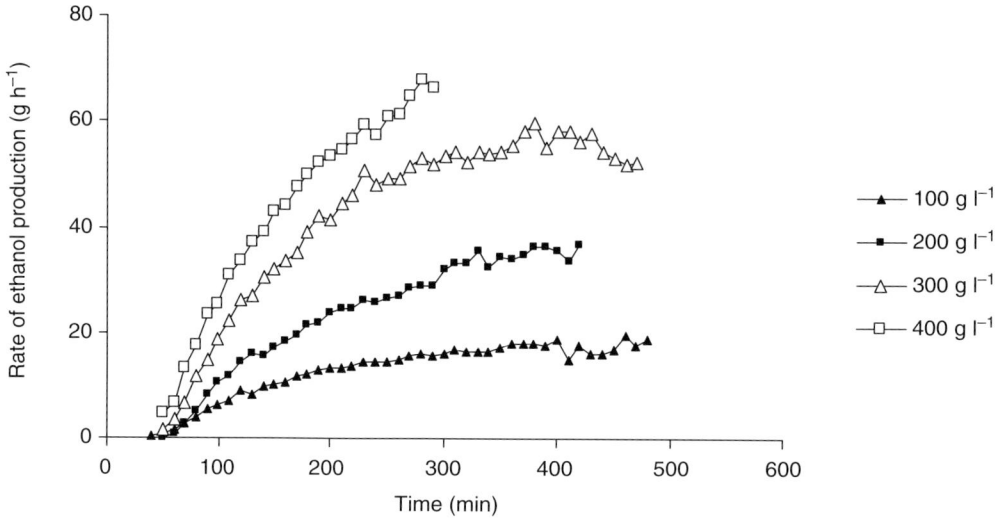

Figure 6.5 Effect of feed glucose concentration on the rate of ethanol production from glucose using extruded yeast pellets. Reproduced from Hayes (1998) with permission.

gas-solid fluidized bed fermentation, carried out close to atmospheric pressure, higher ethanol yields have been obtained under nitrogen than under carbon dioxide (Moebus and Teuber, 1982a; Rottenbacher, 1985).

Increasing the concentration of the glucose feed solution (feed D in Table 6.3), whilst keeping constant both the volumetric feed rate and the concentrations of potassium and magnesium salts, resulted in an increase in the rate of ethanol production (Figure 6.5). At the maximum concentration of $400\,g\,l^{-1}$, the glucose feed rate of $189.7\,g\,h^{-1}$ was in excess of the capacity of the glucose sink and resulted in bed quenching.

A model for fluidized bed fermentation

Material and energy balances

Figure 6.6 shows a schematic diagram of an anaerobic gas-solid fluidized bed fermenter. It is a closed system consisting of a bubbling bed of yeast pellets supported by recirculating inert gas. An aqueous glucose solution is sprayed into the bed and the glucose is fermented to ethanol and carbon dioxide. The off-gas passes in sequence through a water-cooled partial condenser, a gas blower, a gas heater and back to the bed. The only input to the system is the aqueous glucose solution and the outputs are an aqueous ethanol solution and carbon dioxide

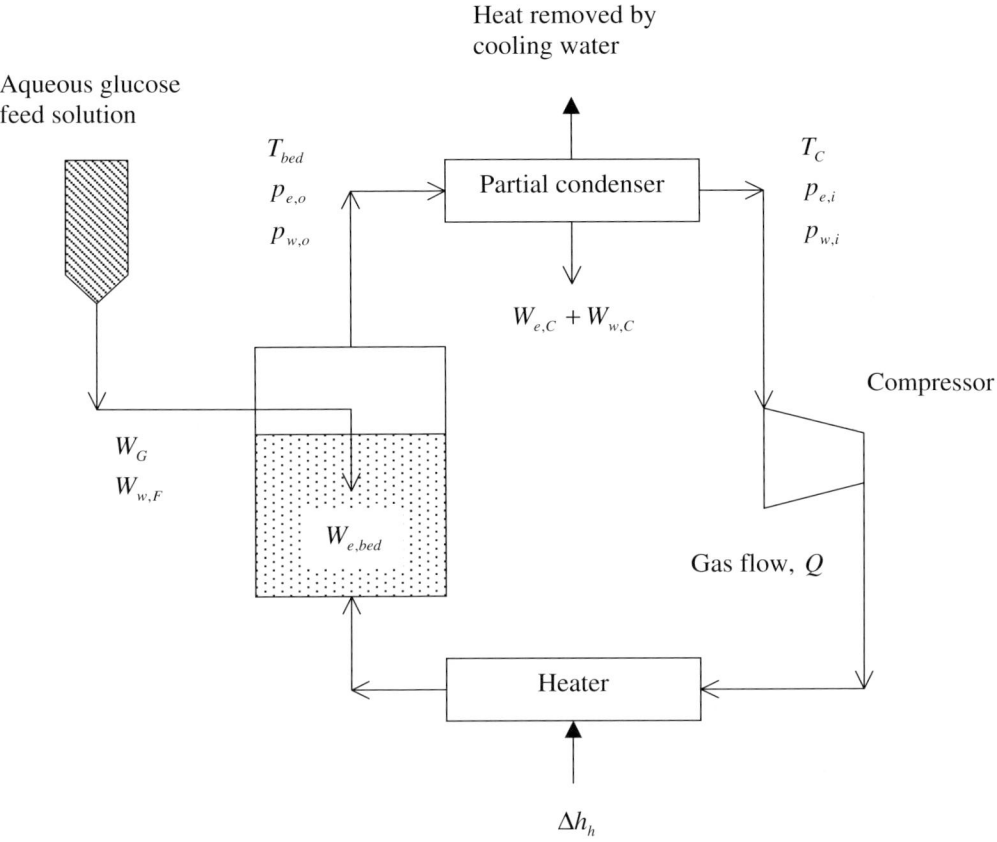

Figure 6.6 Schematic diagram of an anaerobic gas-solid fluidized bed fermenter for ethanol production. Reproduced from Hayes (1998) with permission.

gas which must be vented to maintain the system pressure close to atmospheric. The heater supplies the latent heat to evaporate ethanol and water from the bed which is then removed in the partial condenser to give an aqueous ethanol solution.

The material and energy balances (Beck and Bauer, 1989) assume that the yeast pellets are ideally mixed in the fermenter and that the ethanol and water in the pellets are in equilibrium with the fluidizing gas. This latter assumption is based on two separate ideas: that the yeast-ethanol-water system behaves like an ethanol-water system; and that the inert gas leaving the bed is saturated with water vapour. Rottenbacher (1985) demonstrated that the first of these was reasonable by measuring the adsorption of ethanol vapour from an inert carrier gas (nitrogen) directed upwards through a bed of yeast pellets. Confirmation that the off-gas is saturated with water vapour during yeast growth experiments comes from humidity measurements made by Moebus *et al.* (1981) and Mishra *et al.* (1982). It is important to note that between start-up of the system and equilibrium conditions there is a gain in bed mass due to ethanol accumulation, with the mass of water in the bed remaining unchanged, i.e. the bed does not accumulate ethanol at the expense of water during this period. The following balances assume that equilibrium conditions have been reached.

At equilibrium conditions, the rate at which water enters the bed with the feed $W_{w,F}$ is equal to the rate at which it is removed at the condenser, $W_{w,C}$. The gain in water vapour partial pressure in the fluidizing gas across the bed $(p_{w,o} - p_{w,i})$ depends upon the volumetric flow rate of fluidizing gas Q, the total pressure in the system P, and the density of water vapour ρ_{vap}. Thus

$$W_{w,F} = W_{w,C} \qquad 6.1$$

and

$$W_{w,F} = \frac{\rho_{vap}Q(p_{w,i} - p_{w,o})}{P} \qquad 6.2$$

where $p_{w,o}$ and $p_{w,i}$ are the partial pressures of water vapour in the gas leaving the bed and the condenser respectively. The density of water vapour is determined at the arithmetical average of the mean bed and condenser temperatures, T_{bed} and T_C respectively. Similarly, for ethanol

$$W_{e,bed} = W_{e,C} \qquad 6.3$$

and

$$W_{e,bed} = \frac{\rho_e Q(p_{e,o} - p_{e,i})}{P} \qquad 6.4$$

where $p_{e,o}$ and $p_{e,i}$ are the partial pressures of ethanol vapour in the gas leaving the bed and the condenser respectively.

Assuming the stoichiometric conversion of glucose to ethanol then $W_{e,bed} = 0.511\,W_G$ where W_G is the mass flow rate of glucose into the bed and thus the mass rate of accumulation of carbon dioxide in the system is $0.489\,W_G$. The overall material balance at equilibrium for a system originally charged with inert gas is then

$$W_{w,F} + W_G = W_{w,C} + W_{e,C} + W_{CO_2} \qquad 6.5$$

where the mass inputs to the system are water $W_{w,F}$ and glucose W_G, and the outputs are water $W_{w,C}$, ethanol $W_{e,C}$ and carbon dioxide W_{CO_2}.

At start-up, before any ethanol is generated by fermentation, the initial mass of yeast (water plus yeast dry solids) in the bed is m_y. Any ethanol accumulation in the bed is in excess of m_y so that at equilibrium the total mass in the bed (yeast plus ethanol) is $m_y + m_e^*$. A material balance for the bed mass is then

$$m_y + m_e^* = m_w + m_{DW} + m_e^* \qquad 6.6$$

where m_w is the mass of water in the bed and m_{DW} is the mass of dry yeast solids in the bed.

The overall enthalpy balance at equilibrium is:

$$\Delta h_h + \Delta h_B = \Delta h_w + \Delta h_e + \Delta h_g + \Delta h_L \qquad 6.7$$

where Δh_B represents the heat of reaction generated by the fermentation of glucose and Δh_h is the heat added to the fluidizing gas in the compressor. This balances the enthalpy change of the water in the feed solution Δh_w, of the ethanol produced in the bed Δh_e, of the inert gas Δh_g and the heat lost to the environment Δh_L. Sensible heat is required to heat the water vapour and ethanol in the fluidizing gas leaving the condenser at T_C up to bed temperature T_{bed}. Sensible heat is also required to heat the water in the feed solution from the reservoir temperature T_R up to the bed temperature. The latent heat of vaporisation $h_{fg,w}$ has to be supplied to evaporate water in the feed solution and $h_{fg,e}$ to evaporate ethanol from the bed. Thus

$$\Delta h_w = \left[\rho_{vap} Q \left(\frac{P_{w,i}}{P} \right) c_{p,w} (T_{bed} - T_C) \right] + [W_{w,F} c_{p,w} (T_{bed} - T_R)] \\ + W_{w,F} h_{fg,w} \qquad 6.8$$

and

$$\Delta h_e = \left[\rho_e Q \left(\frac{P_{e,i}}{P} \right) c_{p,e} (T_{bed} - T_C) \right] + W_{e,bed} h_{fg,e} \qquad 6.9$$

where $c_{p,w}$ and $c_{p,e}$ are the heat capacities of water and ethanol respectively. The rate at which heat is supplied to the inert gas leaving the condenser is given by

$$\Delta h_g = \rho_g Q c_{p,g} (T_{bed} - T_C) \qquad 6.10$$

where ρ_g is the density, and $c_{p,g}$ the heat capacity, of the inert gas.

Ethanol-water vapour-liquid equilibria

For non-ideal solutions, such as the ethanol-water system, at the intermediate ethanol concentrations found in the bed and the condensate during anaerobic gas-solid fluidized bed fermentations, Raoult's law (equation 4.2) is inadequate and an activity coefficient γ must be introduced, so that the partial pressures of ethanol p_e and water p_w over an ethanol-water solution are given by equation 6.11

$$\left.\begin{array}{l} p_e = \gamma_e x_e p_e' \\[2ex] p_w = \gamma_w x_w p_w' \end{array}\right\} \qquad\qquad 6.11$$

The methods most generally used for the calculation of activity coefficients at intermediate pressures are the Wilson (1964) and UNIQUAC (Abrams and Prausnitz, 1975) equations. Wilson's equation was used by Sato *et al.* (1985) to predict the composition of the condensate gas stripped from a packed bed fermenter at 30°C, whilst Beck and Bauer (1989) used the UNIQUAC equation, with temperature-dependent parameters given by Kolbe and Gmehling (1985), for their model of an anaerobic gas-solid fluidized bed fermenter at 36°C. In this case it was necessary to go beyond the temperature range of the source data down to 16°C in order to predict the composition of the fluidizing gas leaving the condenser.

The presence of solutes other than ethanol might be expected to reduce the mole fractions of ethanol and water and influence the non-ideality of the ethanol-water system. However, both Williams (1983), who modelled a batch wine fermentation, and Rottenbacher (1985), in ethanol sorption experiments with yeast pellets in a fluidized bed, established that the ethanol-water-yeast system behaves as if the water and ethanol content of the pellets were a simple ethanol-water solution supported by a solid matrix which influences neither mole fractions nor activity coefficients.

The model of Beck and Bauer (1989)

The model of Beck and Bauer (1989) predicts the ethanol productivity, and the ethanol concentrations in the bed and the condensate, assuming equilibrium conditions in an anaerobic gas-solid fluidized bed fermenter using a partial condenser (see Figure 6.6). This model does not predict the build-up of ethanol in the bed nor the increase in the rate of ethanol production at the partial condenser. Rather, it is assumed that this start-up phase is already complete, and that the ethanol concentration in the bed and the rate of ethanol production at the partial condenser have attained steady values. A number of assumptions are made, namely that:

(1) fermentation takes place under anaerobic conditions
(2) there is no substrate limitation and inhibition of fermentation, or restriction of glucose supply caused by agglomeration
(3) there is no mass transfer limitation within the yeast pellets, which implies, first, that the kinetics of ethanol production are those for free cells in submerged culture, second, that the specific rate of ethanol production is at its maximum limited only by ethanol inhibition, and third, that the aqueous ethanol solution in the liquid phase of the pellets is in equilibrium with the surrounding gas
(4) the bed exhibits ideal mixing and that both the assumed bed temperature (36°C) and the ethanol concentration within the pellets are uniform
(5) the ethanol and water vapour in the exhaust gas from the bed is condensed at the partial condenser without a change in the ethanol to water mole ratio; and
(6) the ethanol-water concentration of the recirculated gas is calculated as a function of the vapour pressures at the condenser temperature.

Carbon dioxide build-up in the fermenter is ignored in the model and it is assumed that atmospheric pressure prevails at all points in the system. The partial pressures of ethanol and water vapour over the bed are evaluated at the bed temperature using

$$\left.\begin{array}{l} p_{e,o} = \gamma_{e,bed}\, x_{e,bed}\, p'_e \\[2ex] p_{w,o} = \gamma_{w,bed}\, x_{w,bed}\, p'_w \end{array}\right\} \qquad 6.12$$

and at the exit of the partial condenser, at the condenser temperature, using

$$\left.\begin{array}{l} p_{e,i} = \gamma_{e,C}\, x_{e,C}\, p'_e \\[2ex] p_{w,i} = \gamma_{w,C}\, x_{w,C}\, p'_w \end{array}\right\} \qquad 6.13$$

Beck and Bauer (1989) used an empirical kinetic model (derived by Rottenbacher *et al.* (1985a) for the commercial baker's yeast strain DHW DZ in submerged culture) to define the maximum possible rate of ethanol production in the bed as a function of the dry mass of yeast in the bed and the mole fraction of ethanol in the liquid phase of the bed, thus

$$W_{e,bed\,(max)} = 1.14 m_{DW}\left[1 - \left(\frac{x_{e,bed}}{0.0316}\right)\right] \qquad 6.14$$

The value of $x_{e,bed}$ depends upon the prevailing ethanol-water vapour-liquid equilibria in the bed and the condenser. Assuming that the bed produces ethanol at its maximum possible rate then

$$W_{e,bed\,(max)} = W_{e,C\,(max)} \qquad 6.15$$

Assumption (5) above implies that

$$\frac{p_{e,o}}{p_{w,o}} = \frac{p_{e,i}}{p_{w,i}} \qquad 6.16$$

and that cooling in the condenser is very rapid with very little mass transfer occurring between vapour and condensate. In contrast, if the contact time is sufficient, vapour and condensate will be in equilibrium with one another.

Finally, Beck and Bauer (1989) defined the ethanol productivity of the bed Ψ as

$$\Psi = \frac{W_{e,C}}{V_{bed}} \qquad 6.17$$

where V_{bed} is the bed volume at minimum fluidization. Thus Ψ and the ethanol concentrations in both the bed and the condensate depend upon the bed and condenser temperatures, the dry yeast mass in the bed and the volumetric gas flow rate. The ethanol productivity is limited by ethanol inhibition. For a fixed bed temperature and yeast mass, an increase in gas flow rate Q gives an increase in fermenter productivity and a reduction in the ethanol concentrations in the bed (which in turn reduces ethanol inhibition) and in the condensate. A reduction in condenser temperature has a similar effect. Beck and Bauer (1989) did not report any experimental data to support their theoretical model although the range of reported superficial gas velocities would limit the potential productivity to between 70 and $105\,g\,l^{-1}\,h^{-1}$. This range would undoubtedly be reduced further by diffusional restrictions and agglomeration effects.

The model of Hayes (1998)

In the model of Beck and Bauer (1989) the ethanol concentration in the condensate is always many times greater than that in the bed. In contrast, Hayes (1998) assumed equilibrium condensation of ethanol and water vapour at the partial condenser. Such a condition permits the ethanol concentrations in the bed and the condensate to approach one another as the temperature difference between the bed and the condenser decreases and therefore it predicts higher ethanol concentrations in the bed for the same operating conditions of bed temperature, condenser temperature and fluidizing gas volumetric flow rate. In the Hayes model ethanol inhibition is more severe and this gives lower predicted values for the maximum rate of ethanol production. Hayes assumed that the capacity of the glucose sink is limited only by ethanol inhibition, and that factors affecting the distribution and diffusion of glucose can be ignored, i.e. all yeast cells are actively fermenting glucose. The system was also assumed to be at equilibrium so that the rate of ethanol production at the partial condenser is the steady value and equal to the rate of ethanol production in the bed. Wilson's equation was used rather than UNIQUAC to calculate the partial pressures of ethanol and water vapour.

Modelling the time course of ethanol production

Hayes (1998) proposed that the time course of ethanol production at the condensers is itself a function of the ethanol-water equilibria. Assuming a constant yield of ethanol during the feed period, a constant mass flow rate of glucose gives rise to a constant underlying rate of ethanol production in the bed $W_{e,bed}$. If it is assumed that there is negligible ethanol accumulation as vapour in the gaseous volume of the fermenter system and negligible ethanol accumulation as liquid in the dead volume of the condenser system, then $W_{e,bed}$ is equal to the sum of the rate of ethanol accumulation in the bed and the rate of ethanol production at the condenser $W_{e,C}$

$$W_{e,bed} = \frac{dm_e}{dt} + W_{e,C} \qquad\qquad 6.18$$

where m_e is the mass of ethanol accumulated in the bed. At the start of the feed period $W_{e,C}$ will be zero and the rate of ethanol accumulation in the bed is equal to the underlying rate of ethanol production. As the fermentation progresses, the rate of ethanol accumulation decreases with a corresponding increase in the production rate at the condenser. Ultimately, a point will be reached when these quantities are equal and the mass of ethanol in the bed will have reached its equilibrium value m_e^*. This quantity may be viewed as the capacity of the bed to accumulate ethanol.

It is now possible to define a time constant τ for the bed as the time required for the underlying rate of ethanol production to meet that capacity, thus

$$\tau = \frac{m_e^*}{W_{e,bed}}$$ 6.19

The rate of ethanol accumulation in the bed will depend upon the underlying rate of ethanol production in the bed and the mass of ethanol accumulated in the bed and hence

$$\frac{dm_e}{dt} = W_{e,bed} - \frac{m_e}{\tau}$$ 6.20

Substituting from equation 6.19 gives

$$\frac{dm_e}{dt} = \frac{m_e^* - m_e}{\tau}$$ 6.21

and integrating between the limits $m_e = m_e$ at $t = t$ and $m_e = 0$ at $t = 0$ and combining the result with equation 6.19 gives an expression for the rate of ethanol production at the condensers as a function of time

$$W_{e,C} = W_{e,bed}\left[1 - \exp\left(\frac{-t}{\tau}\right)\right]$$ 6.22

Equation 6.22 assumes that the initial mass of water in the bed remains unchanged over the duration of the fermentation. Hayes (1998) found that agglomeration due to localised wet quenching reduced this initial mass and therefore the bed suffered a net drying rate b. Thus, if a is the initial mass of water in a dehydrating bed at the start of the feed period, the mass of water in the bed at any time during the fermentation is given by

$$m_w = a - bt$$ 6.23

In the dehydrating bed the time constant must be modified by a factor equal to $\left(\frac{a-bt}{a}\right)$. This results in a modified expression for the rate of ethanol production

$$W_{e,C} = W_{e,bed}\left(\frac{a}{a-b\tau}\right)\left[1 - \left(\frac{a-bt}{a}\right)^{\left(\frac{a}{b\tau}-1\right)}\right]$$ 6.24

If equilibrium condensation at the partial condenser is assumed (the Hayes model), then the relevant equations are 6.2, 6.4, 6.12 and 6.13. However, if it is assumed that ethanol and water vapour are condensed without a change in molar ratio (the Beck and Bauer model), it is necessary in addition to satisfy equation 6.16 and to assign hypothetical values of $x_{e,C}$ and $x_{w,C}$ to the calculation of $p_{e,in}$ and $p_{w,in}$. The alternative assumption of equilibrium condensation requires actual values of $x_{e,C}$ and $x_{w,C}$. Hayes (1998) fitted experimental data to equation 6.24 in order to extract the values of τ and $W_{e,bed}$ that gave the highest correlation coefficient.

Comparison of models with experimental data

In Figure 6.7 the time constants predicted by the Hayes and the Beck and Bauer models respectively are compared with the corresponding

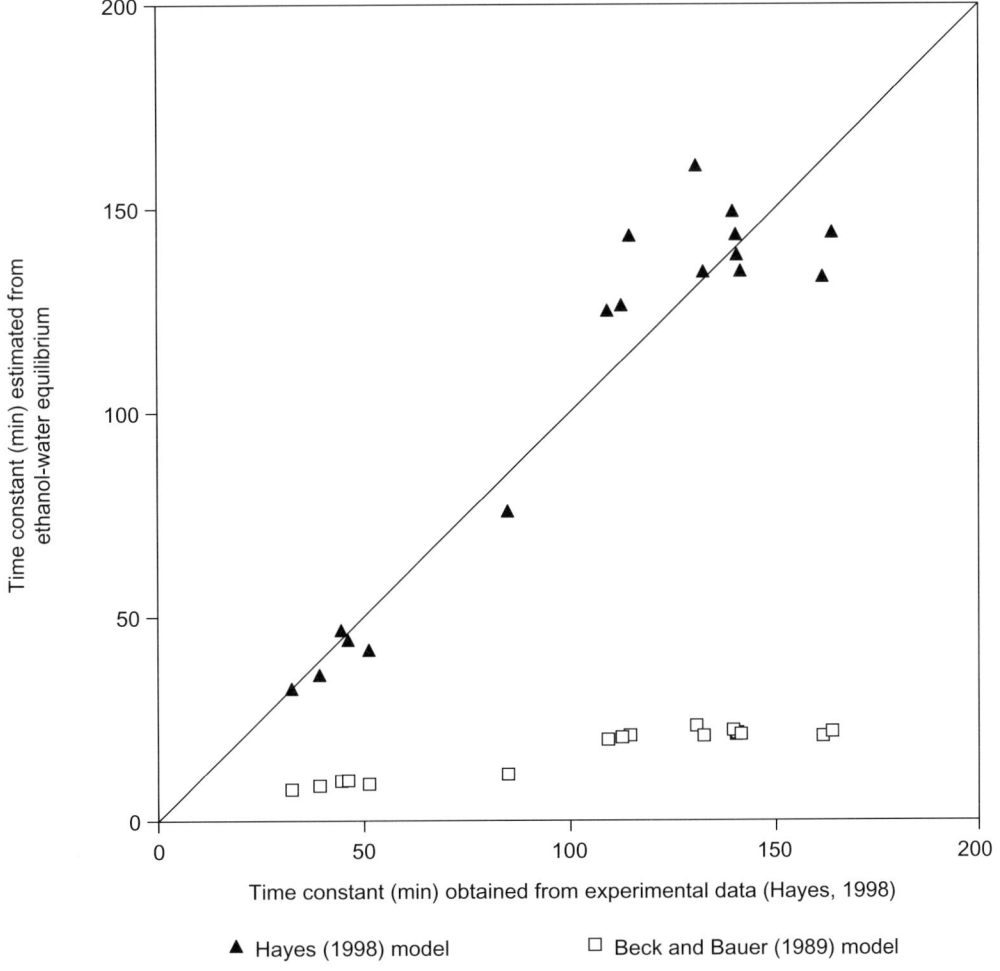

Figure 6.7 Comparison of experimental and predicted time constants. Reproduced from Hayes (1998) with permission.

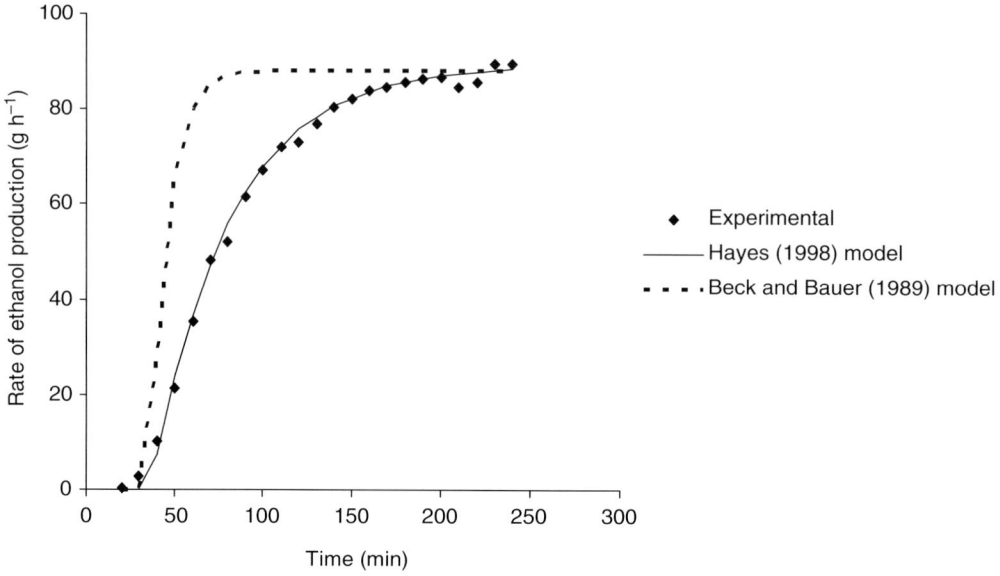

Figure 6.8 A comparison of experimental and predicted rates of ethanol production under anaerobic conditions. Reproduced from Hayes (1998) with permission.

time constants obtained by fitting equation 6.24 to the experimental data obtained by Hayes (1998) from 17 anaerobic fermentations using nominal 1 mm diameter extruded yeast pellets. The good agreement between experiment and the Hayes model supports strongly the assumption of equilibrium condensation of ethanol and water vapour at the partial condenser. The model proposed by Beck and Bauer (1989), which assumes that ethanol and water vapour are condensed at the partial condenser without a change in their molar ratio, leads to considerable underestimation of the experimental time constants and consequently the high ethanol productivities suggested by these authors require considerable downward revision. This is illustrated further by the good agreement between the experimental rate of ethanol production at the condensers for a fermentation under nitrogen and the rate predicted by the Hayes model (Figure 6.8).

Conclusion

Anaerobic, rather than aerobic, fermentation conditions in a gas-solid fluidized bed eliminate the possibility of ethanol reconsumption and result in greater yields of ethanol from glucose. Higher ethanol yields are obtained with nitrogen as the fluidizing gas. Although the highest yields of ethanol are obtained with an enriched medium containing

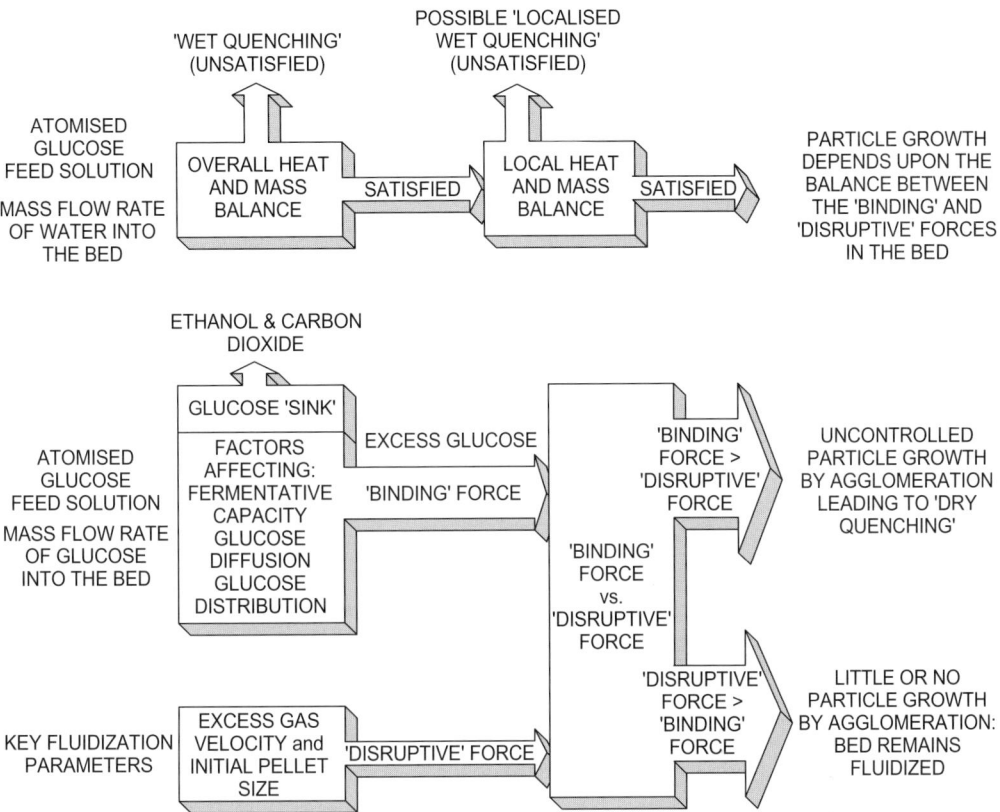

Figure 6.9   Particle growth in anaerobic, gas-solid fluidized bed fermentation. Reproduced from Hayes (1998) with permission.

yeast extract, this can cause early quenching of the bed and therefore an unbuffered glucose solution with added magnesium and potassium salts is the optimum substrate. These salts stimulate fermentation but do not induce bed quenching.

Particle agglomeration and bed quenching are risks associated with adding a liquid to a fluidized bed. A simplified scheme for particle growth in anaerobic, gas-solid fluidized bed fermenters is shown in Figure 6.9. The overall heat and mass balance for the bed must be satisfied so that there is no net accumulation of water. In addition, the mass flow rate of water into the bed with the feed solution has to be sufficiently low to avoid localised wet quenching. Extruded pellets are much less susceptible to localised wet quenching than grated pellets, and enable much higher feed rates to be used. If wet quenching and localised wet quenching are avoided, then the balance between the binding forces and disruptive forces in the bed determines the success of the process. The binding forces increase with glucose concentration and feed rate, whilst the disruptive forces increase with excess gas velocity and initial pellet size. Agitation of the bed also contributes to the breakdown of agglomerates.

Particle growth imposes a limit to how much glucose can be introduced into the bed and therefore limits ethanol productivity. Growth may occur if free glucose is allowed to remain on the surface of the yeast pellets but the binding force is weakened if the feed solution is better distributed. The simultaneous processes of diffusion of glucose into the yeast pellet and fermentation also weaken the binding force. These three factors (distribution, diffusion and fermentation of glucose) constitute the glucose sink. For a given mass of yeast in the bed, an increase in excess gas velocity increases the capacity of the glucose sink by allowing a higher glucose feed rate.

The factors affecting the fermentation of glucose are summarised in Figure 6.10. The ideal capacity of the fermenter is obtained by multiplying the specific rate of ethanol production by the bed dry mass. This ideal value is reduced by distribution and diffusional limitations and may be reduced still further if bed dehydration occurs during fermentation. A bed moisture content of at least 55% (wwb) is required for adequate fermentation of glucose. Below this moisture content rates of ethanol production decline and beds tend to dry quench. Ethanol inhibition reduces the capacity of the glucose sink. By changing the

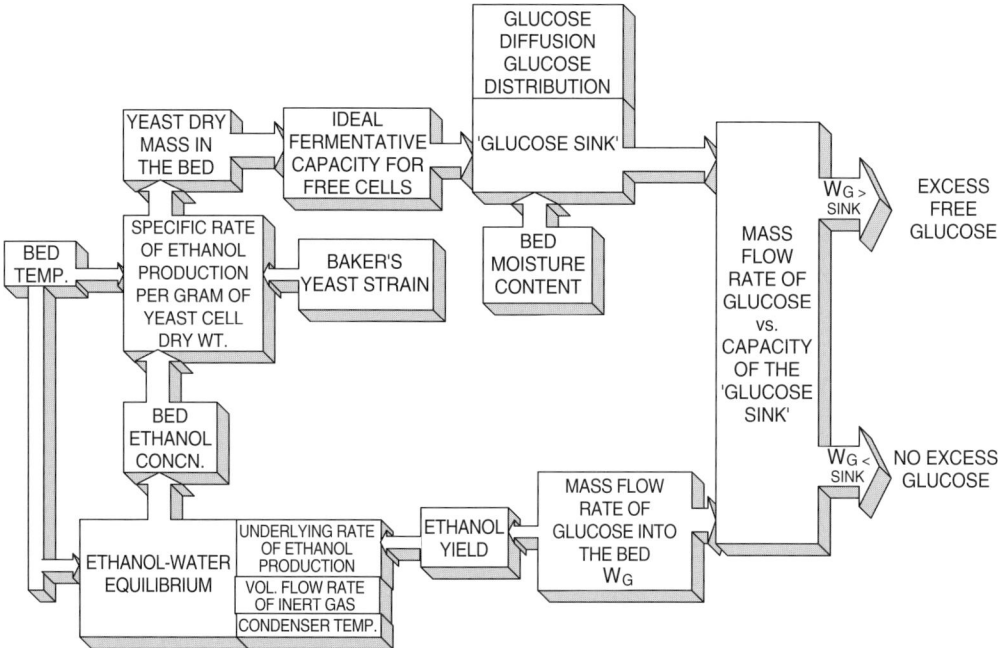

Figure 6.10   Factors affecting the fermentative capacity of the bed and its relationship to the glucose sink. Reproduced from Hayes (1998) with permission.

prevailing ethanol-water equilibrium so that the ethanol concentration in the bed is reduced, higher glucose feed rates are possible. The ethanol-water equilibrium depends upon four key variables: the bed and condenser temperatures, the volumetric flow rate of fluidizing gas, and the underlying rate of ethanol production in the bed. The latter depends upon the mass flow rate of glucose and the ethanol yield. Increasing the bed temperature and gas flow rate, and reducing the condenser temperature and feed glucose concentration, increase the capacity of the glucose sink by reducing ethanol inhibition.

Nomenclature

a	initial mass of water in a dehydrating bed
b	net bed drying rate
$c_{p,e}$	heat capacity of ethanol
$c_{p,g}$	heat capacity of inert gas
$c_{p,w}$	heat capacity of water
$h_{fg,e}$	latent heat of vaporisation of ethanol
$h_{fg,w}$	latent heat of vaporisation of water
m_{DW}	mass of dry yeast solids in the bed
m_e	mass of ethanol in the bed
m_e^*	mass of ethanol in the bed at equilibrium
m_w	mass of water in the bed
m_y	initial mass of yeast in the bed (water + yeast dry solids)
$p_{e,i}$	partial pressure of ethanol vapour leaving the condenser (and recycled to the bed)
$p_{e,o}$	partial pressure of ethanol vapour leaving the bed (and entering the condenser)
p_e'	saturation vapour pressure of ethanol
$p_{w,i}$	partial pressure of water vapour leaving the condenser (and recycled to the bed)
$p_{w,o}$	partial pressure of water vapour leaving the bed (and entering the condenser)
p_w'	saturation vapour pressure of water
P	total system pressure
Q	volumetric flow rate of fluidizing gas
t	time
T_{bed}	mean bed temperature
T_C	condenser temperature
T_R	feed temperature
u_{mf}	minimum fluidizing velocity
V_{bed}	bed volume at minimum fluidization
W_{CO_2}	rate of carbon dioxide production in the bed

$W_{e,bed}$	underlying rate of ethanol production in the bed
$W_{e,C}$	mass flow rate of ethanol from the condenser
W_G	mass flow rate of glucose into the bed
$W_{w,C}$	mass flow rate of water from the condenser
$W_{w,F}$	mass flow rate of water into the bed
x_e	mole fraction of ethanol
x_w	mole fraction of water

Greek symbols

γ_e	ethanol activity coefficient
γ_w	water activity coefficient
Δh_B	heat of reaction (fermentation of glucose)
Δh_e	enthalpy change of ethanol produced in the bed
Δh_g	enthalpy change of inert gas
Δh_h	heat added to the fluidizing gas in the compressor/heater
Δh_L	heat lost to the environment
Δh_w	enthalpy change of water in the feed solution
ρ_e	density of ethanol vapour
ρ_g	density of inert gas
ρ_{vap}	density of water vapour
τ	time constant
Ψ	ethanol productivity

References

Abrams, D.S. and Prausnitz, J.M., Statistical thermodynamics of liquid mixtures: a new expression for the excess Gibbs energy of partly or completely miscible systems, *A. I. Chem. E. J.*, **21** (1975) 116–128.

Akao, T. and Okamoto, Y., Cultivation of microorganisms in an air-solid fluidized bed. Proceedings of the Fourth International Conference on Fluidization, Kashikojima, 1983, 631–637.

Atkinson, B. and Mavituna, F., Biochemical engineering and biotechnology handbook, Macmillan, Basingstoke, 1983.

Bauer, W., Gas/solid fermentation systems – development and application. Proceedings of the 34th Canadian Society of Chemical Engineers Conference, Quebec, 1984, 264–267.

Bauer, W., Der Wirbelschichtfermenter – Entwicklung eines neuen Fermentertyps [Fluid bed fermentation – development of a new fermenter type], *Int. Z. Lebensmittel Tech. Verfahrenstechnik*, **3** (1985) 154–161.

Bauer, W., The use of gas/solid fluidization for biocatalysed reactions: experiments and modelling, in: Ostergaard, K. and Sorensen, A., (eds.), Proceedings of the Fifth International Conference on Fluidization, Elsinore, Denmark, 1986. Engineering Foundation, New York, 619–626.

Bauer, W. and Kirk, H.-G., Synthesis of glutathione with *S. cerevisiae* in solid-state and submerged culture fermentation, *Int. Zeit. Biotech.*, **2** (1985) 130–138.

Bauer, W. and Kunz, B., Solid-state fermentation in food industry, in: Chmiel, H., Hammes, W.P. and Bailey, J.E., (eds.), Biochemical engineering, Gustav Fischer Verlag, Stuttgart, 1987, 228–241.

Bauer, W. and Rottenbacher, L., Regelung der Substratzufuhr eines Gas/Festhoff-Wirbelschichtfermenters über die Bestimmung des Fluidisationsverhaltens [Control of material feed into a gas/solids fluidized bed fermenter on the basis of fluidization behaviour], *Int. Z. Lebensmittel Tech. Verfahrenstechnik*, **35** (1984) 18–23.

Bauer, W. and Rottenbacher, L., Utilisation des lits fluidises gaz-solide en biotechnologie: application a la production d'ethanol par fermentation [Use of gas-solid fluidized beds in biotechnology: application to ethanol production by fermentation], *Entropie*, **124** (1985) 18–23.

Beck, M. and Bauer, W., Energy balance of ethanol production with a gas-solid fluidized bed fermenter, *Bioproc. Eng.*, **4** (1989) 123–128.

Berry, D.R., The biology of yeast, Edward Arnold, London, 1982.

Burrows, S., Baker's yeast, in: Rose, A.H. and Harrison, J.S., (eds.), The yeasts, volume 3, Academic Press, London, 1970, 348–420.

Casey, G.P. and Ingledew, W.M., Ethanol tolerance in yeasts, *Critical Rev. Microbiol.*, **13** (1986) 219–280.

de Bont, J.A.M. and van Ginkel, C.G., Ethylene oxide production by immobilised *Mycobacterium* Py1 in a gas-solid bioreactor, *Enzyme. Microbial. Tech.*, **5** (1983) 55–59.

Egerer, B., Zimmermann, K. and Bauer, W., Flow and fluidization behaviour of yeast in gas/solid fermentation and drying, *I. Chem. E. Symp. Ser.*, **91** (1985) 257–269.

Ellis, S.P., Gray, K.R. and Biddlestone, A.J., Mixing evaluation of a Z-blade mixer developed as a novel solid state bioreactor, *Food Bioprod. Proc.*, **72** (1994) 158–162.

Fiechter, A., Kappeli, O. and Meussdoerffer, F., Batch and continuous culture, in: Rose, A.H. and Harrison, J.S., (eds.), The yeasts, volume 2, 2nd ed., Academic Press, London, 1987, 99–129.

Gancedo, C. and Serrano, R., Energy-yielding metabolism, in: Rose, A.H. and Harrison, J.S., (eds.), The yeasts, volume 3, 2nd ed., Academic Press, London, 1989, 205–259.

Grabowski, S., Mujumdar, A.S., Ramaswamy, H.S. and Strumillo, C., Evaluation of fluidized versus spouted bed drying of baker's yeast, *Drying Tech.*, **15** (1997) 625–634.

Hayes, W.A., Ethanol production from glucose by *Saccharomyces cerevisiae* in an anaerobic gas-solid fluidised bed fermenter, PhD thesis, University of Lincolnshire and Humberside, 1998.

Hou, C.T., Propylene oxide production from propylene by immobilized whole cells of *Methylosinus* sp. CRL 31 in a gas-solid bioreactor, *Appl. Microbiol. Biotech.*, **19** (1984) 1–4.

Ingledew, W.M., Yeasts for production of fuel ethanol, in: Rose, A.H. and Harrison, J.S., (eds.), The yeasts, volume 5, 2nd ed., Academic Press, London, 1993, 245–291.

Jackman, E.A., Industrial alcohol, in: Bu'Lock, J.D. and Kristiansen, B., (eds.), Basic biotechnology, Academic Press, London, 1987, 309–336.

Jones, R.P. and Greenfield, P.F., A review of yeast ionic nutrition, part I. Growth and fermentation requirements, *Process Biochem.*, **19** (1984) 48–62.

Koga, S., Echigo, A. and Nanumora, J., Physical properties of cell water in partially dried *Saccharomyces cerevisiae*, *Biophys. J.*, **6** (1966) 665–674.

Kolbe, B. and Gmehling, J., Thermodynamic properties of ethanol + water, II. Potentials and limits of G^E models. *Fluid Phase Equilibria*, **23** (1985) 227–242.

Maiorella, B., Wilke, C.R. and Blanch, H.W., Alcohol production and recovery, *Adv. Biochem. Eng.*, **4** (1981) 43–92.

Maynard, A.I., The influence of magnesium ions on the growth and metabolism of *Saccharomyces cerevisiae*, PhD thesis, Dundee Institute of Technology, 1993.

Mishra, I.M., El-Temtamy, S.A. and Schugerl, K., Growth of *Saccharomyces cerevisiae* in gaseous fluidized beds, *Eur. J. Appl. Microbiol. Biotechnol.*, **16** (1982) 197–203.

Moebus, O. and Teuber, M., Production of ethanol by solid particles of *Saccharomyces cerevisiae* in a fluidized bed, *Eur. J. Appl. Microbiol. Biotechnol.*, **15** (1982a) 194–197.

Moebus, O. and Teuber, M., Ein neues fermentations-system: Herstellung von ethanol in der wirbelschicht [A new fermentation system: production of ethanol in a fluidized bed], in: Dellweg, H., (ed.), 5th Symp. Technische Mikrobiologie, Berlin, 1982b, 290–297.

Moebus, O. and Teuber, M., Synthesis of glutathione in a gaseous fluidized bed using ethanol producing yeasts as an ATP-regeneration and enzyme system, in: Third European Congress on Biotechnology, volume 1, Verlag Chemie, München, 1984, 553–559.

Moebus, O. and Teuber, M., Direkte Umwandlung von Stärke in Ethanol in einem Gas-Festhoff-Wirbelschichtfermenter mit technischen Amylasen und Bäckerhefe [Direct conversion of starch into ethanol in a gas-solid fluidized bed fermenter with technical amylases and baker's yeast], *Wissenschaft und Umwelt*, **1** (1985) 80–84.

Moebus, O., Teuber, M. and Reuter, H., Growth of *Saccharomyces cerevisiae* in form of solid particles in a gaseous fluidized bed, *Kieler Milchwirtschaftliche Forschungsberichte*, **33** (1981) 3–23.

Moebus, O., Teuber, M. and Reuter, H., Pneumatischer Wirbelschicht-Bioreaktor zur Herstellung von Einzellereiweiß (SCP) oder Ethanol [A gaseous fluidized bed for the production of single cell protein (SCP) or ethanol], *Dechema-Monographien*, **95** (1984) 181–194.

Pamment, N.B., Overall kinetics and mathematical modelling of ethanol inhibition in yeasts, in: van Uden, N., (ed.), Alcohol toxicity in yeasts and bacteria, CRC Press, Boca Raton, Florida, 1989, 1–75.

Pamment, N.B., Dasari, G. and Worth, M.A., Mechanisms of ethanol inhibition in yeasts, in: Yu, P.-L., (ed.), Fermentation technologies: industrial applications, Elsevier Applied Science, London, 1990, 241–246.

Ratledge, C., Biochemistry of growth and metabolism, in: Bu'Lock, J.D. and Kristiansen, B., (eds.), Basic biotechnology, Academic Press, London, 1987.

Rottenbacher, L., Entwicklung und Modellierung eines Gas/Feststoff-Wirbelschichtfermenters für die Erzeugung von Ethanol mit *S. cerevisiae*

[Development and modelling of a gas/solid fluidized bed fermenter for the production of ethanol with *S. cerevisiae*], Dissertation, Technical University of Hamburg-Harburg, 1985.

Rottenbacher, L., Behlau, L. and Bauer, W., The application of infrared gas analyzers for the fast determination of kinetic parameters for ethanol production from glucose, *J. Biotech.*, **2** (1985a) 137–147.

Rottenbacher, L., Schoßler, M. and Bauer, W., Mathematical modelling of alcoholic fermentation in a gas/solid bioreactor – combined effects of solids mixing and non-steady-state kinetics. Proceedings of the First IFAC Symposium on Modelling and Control of Biotechnological Processes, Noordwijkerhout, 1985b, 151–157.

Rottenbacher, L., Schoßler, M. and Bauer, W., Modelling a solid-state fluidized bed fermenter for ethanol production with *S. cerevisiae, Bioproc. Eng.*, **2** (1987) 25–31.

Sato, K., Nakamura, K. and Sato, S., Solid-state ethanol fermentation by means of inert gas circulation, *Biotech. Bioeng.*, **27** (1985) 1312–1319.

Smith, P.G. and Nienow, A.W., Particle growth mechanisms in fluidized bed granulation part 1. The effect of process variables, *Chem. Eng. Sci.*, **38** (1983) 1223–1231.

Smith, P.G., Beers, P.J. and Hayes, W.A., Production of bioethanol in gas-solid fluidised bed fermentation using *Saccharomyces cerevisiae*. Third International Conference on Environmental Impact Assessment, Prague, 1996, 453–457.

Smith, P.G., Beers, P.J. and Hayes, W.A., Ethanol production in a gas-solid fluidised bed fermenter, in Jowitt, R., (ed.), Engineering and food at ICEF 7, Sheffield Academic Press, Sheffield, 1997, B1-B4.

Stear, C.A., Handbook of bread making technology, Elsevier Applied Science, London, 1990.

Strumillo, C., Zbicinski, I. and Liu, X.D., Thermal drying of biomaterials with porous carriers, *Drying Tech.*, **13** (1995), 1447–1462.

Tanaka, M., Kawaide, A. and Matsuno, R., Cultivation of microorganisms in an air-solid fluidized bed fermenter with agitators, *Biotech. Bioeng.*, **28** (1986) 1294–1301.

Thibault, J., Le Duy, A. and Cote, F., Production of ethanol by *Saccharomyces cerevisiae* under high-pressure conditions, *Biotech. Bioeng.*, **30** (1987) 74–80.

van Uden, N., Effects of alcohols on the temperature relations of growth and death in yeasts. in: van Uden, N., (ed.), Alcohol toxicity in yeasts and bacteria, CRC Press, Boca Raton, Florida, 1989, 77–88.

Walker, G.M., The roles of magnesium in biotechnology, *Critical Rev. Biotech.*, **14** (1994) 311–354.

Warren, R.K., Hill, G.A. and Macdonald, D.G., Continuous cell recycle fermentation to produce ethanol, *Food Bioprod. Proc.*, **72** (1994) 149–157.

Williams, L.A., Theory and modelling of ethanol evaporative losses during batch alcoholic fermentation, *Biotech. Bioeng.*, **25** (1983) 1597–1612.

Wilson, G.M., Vapour-liquid equilibrium, XI. A new expression for the excess free energy of mixing, *J. Am. Chem. Soc.*, **86** (1964) 127–130.

Chapter 7
Other Applications of Fluidization

Introduction

The first four chapters in Part Two of this book cover the more significant applications of gas-solid fluidization to the processing of food. However, the unique properties of fluidized beds have led to their use for a much wider range of applications although in some cases the published work is preliminary or entirely experimental in nature and is not necessarily very recent. This chapter is intended to give an idea of the scope of this work without covering any single application in excessive detail. The applications of gas-solid fluidization in the following sections are essentially physical operations most of which involve the addition of heat but which have consequent chemical changes to the food (for example, blanching and cooking) or in the case of sterilisation, microbiological changes to food. In contrast, the most significant use of liquid-solid fluidized beds is as a vehicle for chemical or, more usually, biochemical reactions.

Gas-solid fluidization

Blanching

Blanching is the term for the treatment of fruit and vegetable matter at high temperature in order to inactivate enzymes before the food is frozen, canned or dried. This prevents enzyme activity which might continue even at low temperatures or low moisture contents; for example, blanching is commonly used to preserve the colour of frozen food such as peas. Conventional methods use large quantities of hot water at about 95°C and contribute significantly to the volume of effluent from food plants (Lee, 1975). A further disadvantage is the leaching of salts, sugars and water-soluble vitamins from the food. An alternative to hot water is the use of steam but whilst this may avoid some of the problems, non-uniform heat treatment may be the result because

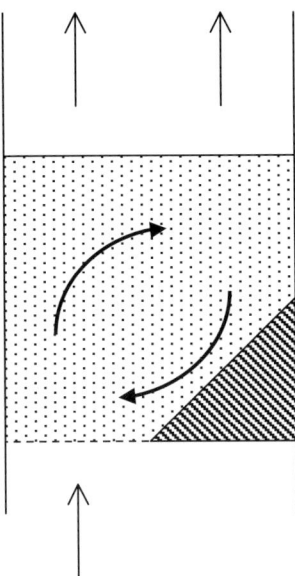

Figure 7.1 Whirling fluidized bed.

food is placed in thick layers on conveyor belts which pass through the steam blancher (Gibert *et al.*, 1979). Fluidized bed blanching, on the other hand, offers rapid and uniform heat treatment and a substantial reduction in the volume of effluent (Gibert *et al.*, 1979).

Rios *et al.* (1978) used a 'whirling' fluidized bed for blanching which consists of a cylindrical bed and a conventional gas distributor plate but with a 45° wedge placed on the plate and covering half its area (Figure 7.1). The wedge induces a cyclical particle motion and high-intensity mixing (Rios *et al.*, 1985) if the superficial gas velocity is above what these authors term the minimum whirling velocity. For particle diameters between 3 and 10 mm, and solid densities between 1000 and 3000 kg m^{-3}, this velocity is approximately constant and equal to 3.5 m s^{-1}. Using a mixture of air and saturated steam as the fluidizing medium, Rios *et al.* (1978) suggested that the blanching time does not depend upon the heat transfer coefficient in the bed if this is over 2900 W m^{-2} K^{-1}; such high values of the heat transfer coefficient were attributed to agitation in the bed which continually renewed the condensate film around a particle and thus reduced its thickness. Consequently the process is governed by the internal thermal resistance of the food. Solids in the whirling bed are perfectly mixed and therefore to control residence time, and improve the quality of the blanched food, Rios *et al.* (1978) proposed using several beds in series in order to eliminate back-mixing. However, they concluded that a very large number of stages would be required and that a better option would be a semi-continuous process with two batch blanchers working alternately and in parallel.

Gibert *et al.* (1979) blanched peas, beans, diced carrot and diced potatoes in a 0.1 m diameter whirling bed, with a gas velocity of 4.5 m s^{-1}, and measured the inactivation of peroxidase as an indication of blanching effectiveness. Plotting enzyme inactivation against time, they concluded that the shape of the curve was similar to that obtained in conventional hot water blanching but that blanching times were shorter. A further development of the concept consisted of a number of whirling cell fluidized beds mounted on a conveyor belt system (Rios *et al.*, 1985) which resulted in a reduction in energy consumption of over 20% compared to water blanching, a significant reduction in the loss of vitamin C and reduced waste water treatment costs.

Roasting

A further example of the exploitation of the heat transfer properties of gas-solid fluidization is found in the roasting of coffee beans which has been reported to result in improved quality compared to the traditional rotating drum method (Rios *et al.*, 1985). Rios and co-workers have also proposed the use of the whirling fluidized bed described above for roasting, in which roasting times were reduced from about 15 minutes in the traditional method to between two and four minutes (Arjona *et al.*, 1980). The optimum bed temperature was between 200 and 260°C; below 200°C the roasting time was increased significantly and above 260°C there was damage to the organoleptic properties of the beans (Rios *et al.*, 1985). A potential disadvantage of the fluidized bed technique is higher levels of pollution due to the rejection of hot exhaust gases. However, coupled with gas recycle of up to 80%, Rios *et al.* (1985) showed that aroma recovery and reduced energy consumption were possible. The recovery of coffee aroma compounds from the exhaust gases can be achieved using a simple water-cooled condenser. Shin and Crouzet (1981) claimed that the most important aroma compounds accounted for the greatest percentage contribution by mass and were the most easily condensed.

Vibrated fluidized beds have also been used for roasting operations including the processing of hazel nuts, almonds, pistachios and desiccated coconut (Weyell, 1997). Murthy *et al.* (1995) found that the products obtained from vibratory units were more uniformly roasted with less overprocessing and superior flavour. They roasted maize, sorghum, wheat and rice at temperatures in the range 300–350°C with residence times of about 45 s. Fluidized bed roasters have been the subject of a number of patents: for roasting cereals, pulses, spices, oilseeds and ready-to-eat snack foods using flue gas (Murthy *et al.*, 2004), green coffee beans (Wilke and Galke, 2001) and candied coffee beans (Winkelmann *et al.*, 2000).

Nagaraju *et al.* (1997) extended the studies of the roasting of coffee beans by using a spouted bed. These authors suggest that fluidized bed roasters are at a disadvantage with small batches and with large particles of about 5 mm in size which tend to develop unstable fluidization and exhibit slugging. They used a 0.15 m diameter bed to roast ellipsoid beans 6.4 by 3.2 mm in size, and with a density of $900 \, \text{kg m}^{-3}$, and found an inverse logarithmic relationship between roasting time in the range 3–11 minutes and temperature in the range 200–320°C. Optimum quality was obtained at 250°C with a batch residence time of 5 minutes. Roast barley, as used in beer production, is another example of a product which is prepared traditionally in rotating drums. Considerably longer roasting times are required than for coffee, typically 2.5 hours at 230°C. In an attempt to develop a more controlled process, Robbins and Fryer (2003) used a spouted bed. They point out that the roasting of barley also involves a degree of drying which is not present with coffee, typically from 15% mass (dry weight basis) to 0%. This means that the logarithmic relationship between batch time and temperature proposed by Nagaraju *et al.* (1997) for coffee does not apply to barley.

Explosion puffing

Expanded or puffed products are formed by the flash evaporation of water due to exposure to high temperature or sudden low pressure (Guraya and Toledo, 1994). The process involves an initial drying stage followed by puffing, for example in a high-temperature fluidized bed (Varnalis *et al.*, 2001). The drying stage creates a partially dried layer at the food surface which seals the surface and reduces the mass transfer of water from the inside of the food particle. Moisture then vaporises upon heating, the pressure increases inside the particle, the walls expand and a porous structure is created (Varnalis *et al.*, 2001). Puffing is essential if a crispy texture is to be achieved in a dried product and requires the development of an internal porous structure but this is possible only if the matrix resists shrinking on drying. Torreggiani *et al.* (1995) prevented tissue collapse in dried and puffed apple cubes by the infusion of gelatinised starch. The apples were first osmo-blanched in corn syrup to prevent enzymatic browning, then treated with starch, before simultaneous drying and puffing in a fluidized bed (Torreggiani and Toledo, 1991). Explosion puffing may be seen as a cheaper, quicker alternative to freeze drying (Varnalis *et al.*, 2004) although the final quality may be a little lower. In many parts of the world fluidized bed puffing is an alternative to the roasting of food in sand (Lai and Cheng, 2004; Chandrasekhar and Chattopadhyay, 1989).

Chandrasekhar and Chattopadhyay (1989) studied heat transfer during the production of puffed rice for use in breakfast cereals. They measured the puffing time, i.e. the time from placing parboiled milled rice grain in the bed to the point of expansion, and found that it decreased with inlet air temperature. The rice grains were initially 7.2 mm in length with an equivalent spherical diameter of 3.2 mm. The expansion ratio (volume of the product divided by the volume of the initial grain) varied between 8.5 and 10 in the range 240–270°C; at lower inlet air temperatures the expansion ratio was below 8.5 and at higher temperatures the rice was discoloured. Puffing times varied between 7 and 9.7 s. Other expanded products which have been produced in a fluidized bed include a fat-free, starch-based snack food containing sweet potato puree and tapioca starch (Guraya and Toledo, 1994); potato cubes (Varnalis *et al.*, 2001); and pre-gelatinised rice flour for use as a bulking or thickening agent (Lai and Cheng, 2004).

Sterilisation

The direct sterilisation of particulate solid foods in a gas-solid fluidized bed was proposed as long ago as 1968 by Lawrence *et al.* (1968) who sterilised wheat flour in steam-air mixtures at the pilot scale. However, Jowitt (1977) described an atmospheric pressure process for the sterilisation of canned foods in which the cans are immersed in a fluidized bed of inert particles. This has a number of advantages compared to the conventional retorting process using pressurised steam or hot water:

(1) the cans are not in direct contact with heating or cooling water and therefore there is no corrosion of either the equipment or the cans
(2) no spoilage of the food occurs due to the ingress of water
(3) there is a reduction in water consumption for the plant
(4) the gas fluidized bed can be heated by combustion products, resulting in improved thermal efficiency.

In a retort which uses condensing steam the heat transfer coefficient to the can is of the order of $5000–10\,000\,\mathrm{W\,m^{-2}\,K^{-1}}$ and therefore the thermal resistance in transferring heat to the centre of the can lies within the food, due largely to the poor thermal conductivities of food materials. Bed-to-surface heat transfer coefficients (see Chapter 2) may be up to $400\,\mathrm{W\,m^{-2}\,K^{-1}}$ and Jowitt suggests that, despite this reduction in heat transfer coefficient by at least an order of magnitude, for foods which heat largely by conduction (for example, baked beans, tomato puree, sago pudding) the dominant thermal resistance still lies within the can. Consequently, his experiments, in which cans were heated in a fluidized bed of 125–180 μm sand at a bed temperature of 121°C, showed that the heat penetration curves for these kinds of foods were very similar for both conventional steam retorting and fluidized bed retorting. Thus the overall processing times were similar.

However, this does not necessarily apply to foods which heat by convection. Jowitt quotes both peas in brine and soup as examples where the process time was doubled in a fluidized bed (22 minutes) compared to a steam-heated retort (11 minutes) for the same total process lethality. However, increasing the fluidized bed temperature by 8 K resulted in almost equal process times and approximately equal retention of the heat-sensitive vitamin thiamine. Following the heating and holding stages of the sterilisation operation, the cans were cooled in a fluidized bed in which heat was removed by cooling water passed through finned tubes immersed in the bed (Jowitt and Thorne, 1971).

Disinfestation of wheat

The disinfestation of wheat by heating in a fluidized bed has been investigated by Evans *et al.* (1983) who used a plug flow bed, 0.8 m in length and 0.2 m wide, capable of handling up to 500 kg h^{-1}. The principle is relatively simple: rapid heating to destroy the insects followed by rapid cooling to a safe handling and storage temperature, the particular advantage of a fluidized bed method being the avoidance of local overheating of the grain which allows inlet air temperatures up to 90°C to be used. The bed was divided by vertical baffles placed along its length so that solids were forced to adopt a zig-zag flow pattern. The grain size was not specified but a superficial gas velocity of 1.6 m s^{-1} was used. Evans *et al.* (1983) measured the mortality of the immature stages of *Rhyzopertha dominica* within wheat and showed that complete disinfestation was obtained with a grain temperature of 65°C. Neither the mean residence time in the bed, which was varied between 180 s and 240 s, nor the longitudinal dispersion of wheat grains in the bed influenced insect mortality. However, disinfestation was successful with temperatures as low as 55°C if the residence time was increased and the grain was allowed to stand for several minutes before cooling. The final stage of the process was cooling, achieved by spraying water onto the grain in a second bed.

Based upon the work of Evans *et al.* (1983), Thorpe (1987) describes a semi-commercial scale continuous-flow fluidized bed disinfestor capable of handling up to 150 t h^{-1} of wheat. He gives a detailed description of the thermodynamic performance of the plant together with a mathematical model of the process which is validated by experimental results obtained from the plant. The plant consists of a single fluidized bed 2 m wide and with an overall length of 9.7 m and divided into four sections for dedusting, heating, pre-cooling and heat recovery, and final cooling respectively. The energy consumption per tonne of grain, at the maximum throughput of 150 th^{-1}, was 4.8 MJ for electricity and 77 MJ for gas. More recently, a disinfestation process using electron beam processing, and what is described as a 'thin' fluidized bed, has been proposed (Anon., 2000). No details of the fluidized bed are given but it is claimed that a low-energy (200–250 keV) electron beam was used successfully with a wide range of particulate foods including sunflower seeds, onion powder, leafy herbs such as parsley and sage, whole peppercorns and various spices, even including whole nutmeg.

Freeze drying

Conventional freeze drying involves freezing of the water in the food, placing the food in a vacuum and then supplying the latent heat of sublimation so that water vapour is formed directly from the solid phase. It is essential that the partial pressure of water vapour in the surrounding air is less than the vapour pressure at the triple point of water (611 Pa). The great advantage of freeze drying is the high quality of the product which is obtained because of the low temperatures used during processing. However, freeze drying is both an expensive and an energy-intensive process, not least because of the need to generate a high vacuum, and consequently it is used only for high-value products. The drawback of conventional freeze drying is the poor heat transfer to, and mass transfer from, the product, which results in long drying times (Yassin and Gibert, 1983).

An alternative possibility is atmospheric pressure freeze drying carried out in a fluidized bed in which pre-frozen food is dried in a bed of fine adsorbent material. The adsorbent plays the same role as the condenser in conventional freeze drying, that is, as a sink for water vapour, and ensures that the vapour pressure is maintained below 611 Pa (Donsi *et al.*, 2001). However, the adsorbent is also the heat transfer medium and, due to the intimate contact between adsorbent particles and the food pieces being dried, the separation of the vapour source and the sink is reduced (Boeh-Ocansey, 1988; Wolff and Gibert, 1991). Thus the heat and mass transfer between the bed and food particles is improved with the water vapour being trapped immediately in the adsorbent particles and the latent heat of sublimation (approximately 2800 kJ kg^{-1}) provided *in situ* by the heat of adsorption; bed temperature and vapour pressure in the bed are uniform (Yassin and Gibert, 1983). Because the heat of sublimation is approximately equal to the heat of adsorption no additional energy is required (Donsi *et al.*, 2001).

The bed particles must be highly adsorbent and the food pieces to be dried as small as possible (Yassin and Gibert, 1983). Boeh-Ocansey (1988) suggested activated carbon, silica gel, activated alumina and molecular sieves as possible adsorbents although one of the major disadvantages of this technique is the need to use food-compatible adsorbents which may exclude the most efficient adsorbent materials (Donsi *et al.*, 2001). Wolff and Gibert (1991) used 160 μm pre-gelatinised corn starch particles and Di Matteo *et al.* (2003) employed bran, corn flour, starch, zeolites and bentonite. Boeh-Ocansey (1986) dried strips of mushroom and beef, carrot discs and cylindrical shrimp pieces, all previously frozen at -35°C, in a fluidized bed of 400 μm diameter activated alumina particles. The minimum fluidizing velocity of the alumina was 0.11 m s^{-1} and the bed was operated at a superficial velocity 50% higher than this. The drying temperature was -10°C, although Rios *et al.* (1985) suggested that drying temperatures as high as 7°C are possible. Following the drying stage, Boeh-Ocansey (1986) regenerated the alumina by heating at 200°C for 30 minutes. Other reported applications include the freeze drying of mushrooms and carrots (Yassin and Gibert, 1983) and shrimps (Donsi *et al.*, 2001).

The advantages of atmospheric pressure fluidized bed freeze drying are the ability to operate continuously (Rios *et al.*, 1985), significant energy savings compared to conventional freeze drying, which were estimated at one-third by Wolff and Gibert (1991), and principally the improved heat and mass transfer. Whilst Di Matteo *et al.* (2003) reported that heat transfer coefficients were more than one order of magnitude higher than in vacuum freeze drying, Wolff and Gibert (1991) considered that the limiting factor was mass transfer. Boeh-Ocansey (1988) found that the ratio of the heat transfer coefficient to the mass transfer coefficient in the fluidized bed process tends to be constant whereas in conventional freeze drying it falls with time. This ratio is between 25 and 40 times greater in a fluidized bed and, as the mass transfer coefficients are similar, heat transfer coefficients are some 20–40 times greater; Boeh-Ocansey (1988) found experimental coefficients in the range 334–468 W m^{-2} K^{-1}.

Liquid-solid fluidization

Bioreactions

The literature on the use of the liquid-solid fluidized bed as a biochemical reactor is vast; much of it is devoted to what may be generally termed fermentation reactions. The principle of the fluidized bed bioreactor, whether based on liquid-solid or on gas-solid fluidization, is one of heterogeneous catalysis. In the case of liquid-solid beds the fluidizing liquid is the substrate for the reaction and the particles usually contain immobilised enzymes or microbial cells which are somehow attached to, or incorporated within, an inert matrix. Epstein (2003) refers to four methods of achieving this. Enzymes or cells may be:

(1) attached to the surface of inert non-porous particles. This results in the growth of a biofilm over the surface of the particle
(2) entrapped within the matrix of a porous particle
(3) encapsulated within a semi-permeable membrane; or
(4) encapsulated within a self-aggregated floc.

The advantages of fluidized beds compared to fixed bed reactors are improved heat and mass transfer, an easier passage through the bed for foreign material and unwanted micro-organisms, a lower pressure drop, better control of the thickness of the biofilm and the easy removal of particles from the bed in order to remove excess biomass (Epstein, 2003).

Calcium alginate beads are a commonly used method of immobilising enzymes (Roy *et al.*, 2004) and their use has been reported for the immobilisation of pullulanase and glucoamylase for the hydrolysis of potato starch (Roy and Gupta, 2004) and the immobilisation of dextransucrase for the synthesis of defined isomalto-oligosaccharides with pre-biotic activity (Berensmeier *et al.*, 2004). In the latter work, the enzyme/alginate beads needed to be weighted with inert silica flour in order to increase the particle density to facilitate its use in the high-density sugar solutions which were used as the reaction substrate. Other particles whose use has been reported are: glucanotransferase immobilised into controlled-pore silica particles for the production of cyclodextrins from maltodextrin (Tardioli *et al.*, 2000); alpha-galactosidase immobilised in polyacrylamide gel for the hydrolysis of oligosaccharides in soymilk (Thippeswamy and Mulimani, 2002); and cellulose beads containing Lactozym™, a commercial enzyme preparation, for the hydrolysis of lactose in whey and whole milk (Roy and Gupta, 2003).

A similar range of immobilisation techniques is reported for whole cells. These include *Candida guilliermondii* cells immobilized on porous glass for the production of Xylitol from sugar cane bagasse (Santos *et al.*, 2005); *Bacillus circulans* cells immobilized in agar gel for the preparation of cyclodextrins from starch (Vassileva *et al.*, 2003); and the fermentation of glucose-xylose mixtures to ethanol using recombinant *Zymomonas mobilis* entrapped in kappa-carrageenan beads (Krishnan *et al.*, 2000).

Fluidized bed reactors employing immobilised cell technology have generated considerable interest in the brewing industry. Donnelly *et al.*

(1999) used yeast immobilised on porous glass beads for continuous beer production and Tata *et al.* (1999), using a similar immobilisation technique, reported that fermentation in a fluidized bed reactor took only half the time required for a conventional batch process. Fluidized beds have been reported to give better performance than fixed bed reactors (Wackerbauer *et al.*, 2003).

Many fluidized reactors are three-phase systems in which gas bubbles, solid particles and the fluidizing liquid are all present. Epstein (2003) points out that such reactors include not only aerobic reactors in which air is bubbled into the bed in order for the reaction to proceed, and in which the gas is also used to help maintain particles in suspension, but also anaerobic reactors in which gas is produced in the course of the reaction. The hydrodynamics of gas lift bioreactors have been reviewed by Petersen and Margaritis (2001). A second significant application of biofilm fluidized bed bioreactors is in the treatment of waste water; the treatment of industrial effluents, including food industry waste water, has been reviewed by Nicolella *et al.* (2000) and by Cooper and Atkinson (1981). Recent studies have included those by Borja and co-workers who investigated the use of anaerobic digestion for the bioremediation of waste water from sunflower meal (Borja *et al.*, 2002) and from the manufacture of protein isolates from chickpea meal (Borja *et al.*, 2004), in each case using magnesium silicate as a support for the immobilised bacteria.

In addition to conventional fluidized bed bioreactors, a variety of fluidized bed geometries are in use including draught tubes to generate internal circulation of liquid, external recirculation of the fluidizing substrate and the use of magnetic bed particles fluidized within an externally generated magnetic field to stabilise fluidization. Bahar and Celebi (2000) report the use of such a bed which employed magnetic polystyrene particles containing immobilized glucoamylase for the hydrolysis of maltose. The exterior of the bioreactor was surrounded by a coaxial solenoid to generate magnetic field strengths between 14 and 40 mT. Finally, liquid-solid fluidized beds can be used as ion exchange columns. Lan *et al.* (2002) report the use of a circulating fluidized bed ion exchange extraction system for the continuous recovery of protein from cheese whey and Sosa *et al.* (2000) recovered lactic acid from fermentation broths.

Sterilisation

Aseptic processing is used to achieve commercial sterility in a continuous flow of liquid or semi-liquid food by heating the food to a suitable temperature before placing it in previously sterilised packaging. The sterilisation of liquids containing particulates presents a number of difficulties because of the variation in residence time between the liquid and solids and between particulates of different sizes. Consequently, considerable variations in the heat treatment or process lethality are likely. One method of solving this problem is to sterilise the liquid and solids separately. Sawada and Merson (1986) proposed that a batch liquid fluidized bed could be used to sterilise the particulates by fluidizing with pressurised hot water, and therefore achieving temperatures above 100°C, followed by cooling with cold sterile water. Hot water could be recycled to reduce energy usage. They reported a mathematical simulation of a 1 m diameter bed used to sterilise potato and beef particles between 1.27 and 2.5 cm diameter. No experimental data were presented but the authors concluded that a practical system could be designed and operated.

Ultrafiltration and reverse osmosis

In ultrafiltration and reverse osmosis, in which solutions are concentrated by allowing the solvent to permeate a semi-permeable membrane, the permeate flux (i.e. the flow of permeate or solvent per unit time, per unit membrane area) declines continuously during operation, although not at a constant rate. Probably the most important contribution to flux decline is the formation of a concentration polarisation layer. As solvent passes through the membrane, the solute molecules which are unable to pass through become concentrated next to the membrane surface. Consequently, the efficiency of separation decreases as this layer of concentrated solution accumulates. The layer is established within the first few seconds of operation and is an inevitable consequence of the separation of solvent and solute.

This phenomenon is usually countered by maximising the degree of turbulence in the feed stream which has the effect of reducing the thickness of the concentration polarisation layer. Turbulent flow can be achieved by careful design of the channel dimensions through which the feed stream flows and by maintaining an adequate feed velocity, or by placing turbulence promoters such as spheres or rods in the flow channel. Fluidized bed particles can be used as turbulence promoters in ultrafiltration and reverse osmosis equipment if the membranes are strong enough to resist the impact of the particles. Rios *et al.* (1985) fluidized 3 mm stainless steel spheres in a vertically aligned tubular alumina membrane of diameter 0.033 m and height 0.65 m. Using this arrangement for the ultrafiltration of milk at 1.5 bar, with an optimum bed voidage of 0.68, these workers found that the permeate flux was 'considerably higher' compared to that without particles present but did not quantify the improvement. The bed was operated at low fluidizing velocities in order to prevent elutriation, therefore suggesting that this approach to the problem of concentration polarisation is applicable only to cases where the throughput or capacity is limited. In earlier work, de Boer *et al.* (1980), in the concentration of cheese whey, used glass beads which caused little damage to conventional membranes if the bead diameter was below 3 mm. They reported that the permeate flux using fluidization was almost equal to that without the presence of particles but at a velocity about 30 times lower. However, with reverse osmosis the fouling layer was not reduced sufficiently with fluidization and, therefore, a lower permeate flux resulted.

Other applications

Other applications of liquid-solid fluidized beds that have been suggested or put into practice include the leaching of vegetable oils from seeds (Epstein, 2003), the freeze concentration of solutions (Rios *et al.*, 1985) and osmotic drying (Marouzé *et al.*, 2001). Fluidization is also the basis of the hydraulic transport of vegetables (McKay *et al.*, 1987). Three-phase fluidized beds have been employed for the fermentation of cocoa beans (Jacquet *et al.*, 1981; Rios *et al.*, 1985).

References

Anon., Disinfection technique could have processors beaming, *Food Quality*, **7** (2000) 36–38, 40.

Arjona, J.L., Rios, G.M. and Gibert, H., Two new techniques for quick roasting of coffee, *Lebensmittel Wissenschaft Technologie*, **13** (1980) 285–290.

Bahar, T. and Celebi, S.S., Performance of immobilized glucoamylase in a magnetically stabilized fluidized bed reactor (MSFBR), *Enzyme Microbial Tech.*, **26** (2000) 28–33.

Berensmeier, S., Ergezinger, M., Bohnet, M. and Buchholz, K., Design of immobilised dextransucrase for fluidised bed application, *J. Biotech.*, **114** (2004) 255–267.

Boeh-Ocansey, O., Low temperature fluidized-bed drying of mushroom, carrot, beef and shrimp samples, *Acta Alimentaria*, **15** (2) (1986) 79–92.

Boeh-Ocansey O., Freeze-drying in a fluidized-bed atmospheric dryer and in a vacuum dryer: evaluation of external transfer coefficients, *J. Food Eng.*, **7** (1988) 127–146.

Borja, R., Gonzalez, E., Raposo, F., Millan, F. and Martin, A., Kinetic analysis of the psychrophilic anaerobic digestion of wastewater derived from the production of proteins from extracted sunflower flour, *J. Agric. Food Chem.*, **50** (2002) 4628–4633.

Borja, R., Rincon, B., Raposo, F., Dominguez, J.R., Millan, F. and Martin, A., Mesophilic anaerobic digestion in a fluidised-bed reactor of wastewater from the production of protein isolates from chickpea flour, *Process Biochem.*, **39** (2004) 1913–1921.

Chandrasekhar, P.R. and Chattopadhyay, P.K., Heat transfer during fluidized bed puffing of rice grains, *J. Food Proc. Eng.*, **11** (1989) 147–157.

Cooper, P.F. and Atkinson, B., (eds.), Biological fluidised bed treatment of water and wastewater, Ellis Horwood, Chichester, 1981.

de Boer, R., Zomerman, J.J., Hiddink, J., Aufderheyde, J., van Swaay, W.P.M., and Smolders, C.A., Fluidized beds as turbulence promoters in the concentration of food liquids by reverse osmosis, *J. Food Sci.*, **45** (1980) 1522–1528.

Di Matteo, P., Donsi, G. and Ferrari, G., The role of heat and mass transfer phenomena in atmospheric freeze-drying of foods in a fluidised bed, *J. Food. Eng.*, **59** (2003) 267–275.

Donnelly, D., Bergin, J., Gardiner, S. and Cahill, G., Kinetics of sugar metabolism in a fluidized bed bioreactor for beer production, Technical Quarterly, *Master Brewers' Assoc. Am.*, **36** (2) (1999) 183–185.

Donsi, G., Ferrari, G. and Di Matteo, P., Utilization of combined processes in freeze-drying of shrimps, *Food Bioprod. Proc.*, **79** (2001) 152–159.

Epstein, N., Applications of liquid-solid fluidization, *Int. J. Chem. Reactor Eng.*, **1** (2003) 1–16.

Evans, D.E., Thorpe, G.R. and Dermott, T., The disinfestation of wheat in a continuous-flow fluidized bed, *J. Stored Prod. Res.*, **19** (1983) 125–137.

Gibert, H., Baxerres, J.L. and Kim, H., Blanching time in a fluidized bed, in: Linko, P., (ed.), Food process engineering, volume 1, food processing systems. Second International Congress on Engineering and Food, Helsinki 1979, Applied Science Publishers, London, 1979, 75–85.

Guraya, H.S. and Toledo, R.T., Volume expansion during hot air puffing of a fat-free starch-based snack, *J. Food Sci.*, **59** (1994) 641–643.

Jacquet, M., Vincent, J.C., Rios, G.M. and Gibert, H., Fermentation of cocoa beans in a triple phase fluidized bed, *Café Cacao The*, **25** (1981) 45–54.

Jowitt, R., Heat transfer in some food processing applications of fluidisation, *Chem. Engnr.*, **November** (1977) 779–782.

Jowitt, R. and Thorne, S.N., Evaluates variables in fluidized-bed retorting, *Food Eng.*, **43** (11) (1971) 60–64.

Krishnan, M.S., Blanco, M., Shattuck, C.K., Nghiem, N.P. and Davison, B.H., Ethanol production from glucose and xylose by immobilized *Zymomonas mobilis* CP4(pZB5), *Appl. Biochem. Biotech.*, **84** (2000) 525–542.

Lai, H.M. and Cheng, H.H., Properties of pregelatinized rice flour made by hot air or gum puffing, *Int. J. Food Sci. Tech.*, **39** (2004) 201–212.

Lan, Q., Bassi. A., Zhu, J.X. and Margaritis, A., Continuous protein recovery from whey using liquid-solid circulating fluidized bed ion-exchange extraction, *Biotech. Bioeng.*, **78** (2002) 157–163.

Lawrence, B., Olsen, R.A. and Liepa, A.L., Fluidized heating process for microbial destruction in wheat flour, *Chem. Eng. Prog. Symp. Series*, **64** (86) (1968) 77–84.

Lee, C.Y., New blanching techniques, *Korean J. Food Sci. Tech.*, **7** (1975) 100–106.

Marouzé, C., Giroux, F., Collignan, A. and Rivier, M., Equipment design for osmotic treatments, *J. Food Eng.*, **49** (2001) 207–221.

McKay, G., Murphy, W.R. and Jodieri-Dabbaghzadeh, S., Fluidisation and hydraulic transport of carrot pieces, *J. Food Eng.*, **6** (1987) 377–399.

Murthy, K.V., Srinivasa Rao, P.N. and Ramesh, T., Continuous vibro fluid bed roaster for breakfast cereals, *Indian Food Industry*, **14** (1995) 35–38.

Murthy, K.V., Jayaprakashan, S.T.G. and Murugan, R.E, Continuous vibro fluidized bed roaster using flue gas, United States Patent, US 6810794 B2, 2004.

Nagaraju, V.D., Murthy, C.T., Ramalakshmi, K. and Srinivasa Rao, P.N., Studies on roasting of coffee beans in a spouted bed, *J. Food Eng.*, **31** (1997) 263–270.

Nicolella, C., van Loosdrecht, M.C.M. and Heijnen, J.J., Wastewater treatment with particulate biofilm reactors, *J. Biotech.*, **80** (2000) 1–33.

Petersen, E.E. and Margaritis, A., Hydrodynamic and mass transfer characteristics of three-phase gaslift bioreactor systems, *Critical Rev. Biotech.*, **21** (2001) 233–294.

Rios, G.M., Gibert, H. and Baxerres, J.L., Factors influencing the extent of enzyme inactivation during fluidized bed blanching of peas, *Lebensmittel Wissenschaft Technologie*, **11** (1978) 176–180.

Rios, G.M., Gibert, H. and Baxerres, J.L., Potential applications of fluidisation to food preservation, in: Thorne, S., (ed.), Developments in food preservation, volume 3, Elsevier, London, 1985, 273–304.

Robbins, P.T. and Fryer, P.J., The spouted-bed roasting of barley: development of a predictive model for moisture and temperature, *J. Food Eng.*, **59** (2003) 199–208.

Roy, I. and Gupta, M.N., Lactose hydrolysis by Lactozym™ immobilized on cellulose beads in batch and fluidized bed modes., *Proc. Biochem.*, **39** (2003) 325–332.

Roy, I. and Gupta, M.N., Hydrolysis of starch by a mixture of glucoamylase and pullulanase entrapped individually in calcium alginate beads, *Enzyme Microbial Tech.*, **34** (2004) 26–32.

Roy, I., Jain, S., Teotia, S. and Gupta, M.N., Evaluation of microbeads of calcium alginate as a fluidized bed medium for affinity chromatography of *Aspergillus niger* pectinase, *Biotech. Prog.*, **20** (2004) 1490–1495.

Santos, J.C., Silva, S.S., Mussatto, S.I., Carvalho, W. and Cunha, M.A.A., Immobilized cells cultivated in semi-continuous mode in a fluidized bed reactor for xylitol production from sugarcane bagasse, *World J. Microbiol. Biotech.*, **21** (2005) 531–535.

Sawada, H. and Merson, R.L., Estimation of process conditions for bulk sterilization of particulate foods in water-fluidized beds, in: Le Maguer, M. and Jelen, P. (eds.), Food engineering and process applications, volume 1: transport phenomena. International Congress on Engineering and Food, Edmonton 1985, Elsevier, London, 1986, 569–581.

Shin, H.K. and Crouzet, J., Recovery of coffee aroma during roasting I. Trapping emitted gases by condensation and absorption, *Café Cacao The*, **25** (1981) 127–136.

Sosa, A.V., Ochoa, J. and Perotti, N.I., Modeling of direct recovery of lactic acid from whole broths by ion exchange adsorption, *Bioseparation*, **9** (2000) 283–289.

Tardioli, P.W., Zanin, G.M. and de Moraes, F.F., Production of cyclodextrins in a fluidized-bed reactor using cyclodextrin-glycosyl-transferase, *Appl. Biochem. Biotech.*, **84** (2000) 1003–1020.

Tata, M., Bower, P., Bromberg, S. et al., Immobilized yeast bioreactor systems for continuous beer fermentation, *Biotech. Prog.*, **15** (1999) 105–113.

Thippeswamy, S. and Mulimani, V.H., Enzymic degradation of raffinose family oligosaccharides in soymilk by immobilized alpha-galactosidase from *Gibberella fujikuroi*, *Proc. Biochem.*, **38** (2002) 635–640.

Thorpe, G.R., The thermodynamic performance of a continuous-flow fluidized bed grain disinfestor and drier, *J. Agric. Eng. Res.*, **37** (1987) 27–41.

Torreggiani, D. and Toledo, R.T., Simultaneous puffing and dehydration of osmotically pre-treated apples cubes in a high temperature fluidized bed (HTFB) drier, in: Mujumdar, A.S. and Filkova, I., (eds.), Drying '91. Proceedings of the 7th International Drying Symposium, Prague 1990, Elsevier, Amsterdam, 1991, 483–488.

Torreggiani, D., Toledo, R.T. and Bertolo, G., Optimization of vapor induced puffing in apple dehydration, *J. Food Sci.*, **60** (1995) 181–185, 194.

Varnalis, A.I., Brennan, J.G. and MacDougall, D.B., A proposed mechanism of high-temperature puffing of potato, Part I. The influence of blanching and drying conditions on the volume of puffed cubes, *J. Food Eng.*, **48** (2001) 361–367.

Varnalis, A.I., Brennan, J.G., MacDougall, D.B. and Gilmour, S.G., Optimisation of high temperature puffing of potato cubes using response surface methodology, *J. Food Eng.*, **61** (2004) 153–163.

Vassileva, A., Burhan, N., Beschkov, V. et al., Cyclodextrin glucanotransferase production by free and agar gel immobilized cells of *Bacillus circulans* ATCC 21783, *Process Biochem.*, **38** (2003) 1585–1591.

Wackerbauer, K., Ludwig, A., Moehle, J. and Legrand, J., Measures to improve long term stability of main fermentation with immobilised yeast, *Monatsschrift Brauwissenschaft*, **56** (2003) 210–215.

Weyell, M., Roasting with vibrations, *Süßwaren*, **41** (3) (1997) 29–30.

Wilke, T. and Galke, P., Small roaster for roasting of green coffee beans and similar products, German Patent, DE 19941036 A1, 2001.

Winkelmann, M., Roebert, L., Arndt, T. et al., Process for candying of coffee beans, German Patent, DE 19902786 C1, 2000.

Wolff, E. and Gibert, H., Freeze-drying under vacuum and in an adsorbing fluidized bed: influence of operating pressure on drying kinetics, in: Mujumdar, A.S. and Filkova, I., (eds.), Drying '91. Proceedings of the 7th International Drying Symposium, Prague 1990, Elsevier, Amsterdam, 1991, 237–246.

Yassin, K.E. and Gibert, H., Atmospheric freeze-drying of foods in a fluidized bed of a finely divided adsorbent, in: McLoughlin, J.V. and McKenna, B.M., (eds.), The production, preservation and processing of food. Proceedings of the 6th International Congress of Food Science and Technology, Dublin 1983, International Congress of Food Science and Technology, Boole, 1983, 208–209.

Index